THE PHYSICS OF BASEBALL

THE PHYSICS
OF BASEBALL

ROBERT KEMP ADAIR

Sterling Professor of Physics
Yale University
and
Physicist to the National League 1987–1989

 HarperPerennial

A Division of HarperCollins*Publishers*

FIRST EDITION

Designed by Helene Berinsky

Library of Congress Cataloging-in-Publication Data

Adair, Robert Kemp.
  The physics of baseball/Robert Kemp Adair.
    p. cm.
  ISBN 0-06-055188-7
  ISBN 0-06-096461-8 (pbk.)
  1. Physics. 2. Baseball. 3. Force and energy. I. Title.
QC26.A23   1990
796.357'01'53—dc20   89-45623

90 91 92 93 94 **MPCCC** 10 9 8 7 6 5 4 3 2 1
    92 93 94 **MPCCC** 10 9 8   (pbk.)

To the memory of my grandfather, Theodore Wiegman (1876–1953), who often sat with a small boy behind third base watching 3-I League baseball in the old Fort Wayne ball park in the 1930s, and to the memory of my son, James Cleland Adair (1957–1978), graceful first baseman on his championship Little League team.

CONTENTS

LIST OF FIGURES

PREFACE

Late in the summer of 1987, Bart Giamatti, then President of the National Baseball League and later Commissioner of Baseball, an old friend and colleague of mine from his days as Professor of English and then President of Yale, asked me to advise him on the (few) elements of baseball that might be best addressed by a physicist. I told Bart that I would be delighted to do so—and that I expected I would have so much fun at such a job that I would find it incorrect to accept any payment as consultant. Bart, ever the English Professor and sensitive to words, responded by appointing me as "Physicist to the National League"—a title that absolutely charmed the ten-year-old boy who I hope will always be a part of me.

When I considered the few questions Bart posed for me, it became increasingly clear that to answer any question properly, I had better understand almost all baseball as best I could. Hence, more as a delightful hobby than as an obligation, I attempted to describe quantitatively as much of the action of baseball as I could in a report to Bart Giamatti as League President. Bart then suggested that I publish the report, suitably expanded, as a book. I was pleased that Bart saw the final manuscript before his death and that he liked it.

Written for fun, and originally for Bart Giamatti, this book is not meant as a scholarly compendium of research on baseball—though I have borrowed extensively from the work of others. Moreover, it is not meant as the definitive treatise on the aspects of baseball that are considered. I have not hesitated to make best estimates on matters that I do not understand as well as I would like, and, though I hope I have made no egregious blunders, the physics of baseball is not trivial and I may have slipped somewhere.

Designed for those interested in baseball, not in the simple principles of physics, I have slighted descriptions of the details of the calculations to which I refer. These calculations were usually conducted by simple BASIC programs on a personal computer. The few formulas which are included, segregated as Technical Notes at the end of each chapter, are introduced only to provide a succinct description of the models I used for those who might be especially interested.

Robert Kemp Adair
Hamden, Connecticut

THE PHYSICS OF BASEBALL

MODELS AND THEIR LIMITATIONS

A small, but interesting, portion of baseball can be understood on the basis of physical principles. The flight of balls, the liveliness of balls, the structure of bats, and the character of the collision of balls and bats are a natural province of physics and physicists.

In his analysis of a real system, a physicist constructs a well-defined model of the system and addresses the model. The system we address here is baseball. In view of the successes of physical analyses in understanding arcane features of nature—such as the properties of the elementary particles and fundamental forces that define our universe (my own field of research) and the character of that universe in the first few minutes of creation—it may seem curious that the physics of baseball is not at all under control. We cannot calculate from first principles the character of the collision of an ash bat with a sphere made up of layers of different tightly wound yarns nor do we have any precise understanding of the effect of the airstream on the flight of that sphere with its curious yin-yang pattern of stitches. What we can do is construct plausible models of those interactions that play a part in baseball and do not violate basic principles of mechanics. These basic principles—such as the laws of the conservation of energy and momentum—severely constrain such models, though they do not completely define them. It is necessary for the models to touch the results of observations—or the results of the controlled observations called experiments—at some points so that the model can be more precisely defined and used to interpolate known results or to extrapolate such results. Baseball, albeit rich in anecdote, has not been subject to extensive quantitative studies of its mechanics, hence, models of baseball are not as well founded as they might be.[1]

However connected with experience, model and system—map and territory—are not the same. The physicist can usually reach precise conclusions about the character of the model. If the model is well chosen so as to represent the salient points of the real system adequately, conclusions derived from an analysis of the model can apply to the system to a useful degree. Conversely, conclusions—although drawn in a logically impeccable manner from premises defined precisely by the model—may not apply to the system because the model is a poor map of the system.

Hence, in order to consider the physics of baseball, I had to construct an ideal baseball game which I could analyze that would be sufficiently close to the real game so that the results of the analysis would be useful. The analysis was easy; the modeling was not. I found that neither my experience playing baseball (poorly) as a youth nor my observations of play by those better fitted for the game than I ideally prepared me for the task of constructing an adequate model of the game. However, with the aid of seminal work by physicists Lyman Briggs, Paul Kirkpatrick,[2] and others, and with help from discussions with other students of the game, such as my long-time associate R. C. Larsen, I believe I have been able to arrive at a sufficient understanding of baseball so that some

---

[1]Curiously, better analyses have been made of golf, probably because there are economic advantages to the support of research by manufacturers who might make and sell better balls and better clubs. Baseballs and (largely) baseball bats are made to specifications set down by major league officials and are less subject to manufacturers' improvements.

[2]Papers by Briggs and Kirkpatrick were published in the *American Journal of Physics;* Briggs's paper in 1959 and Kirkpatrick's paper in 1963.

interesting conclusions from analyses of my construction of the game are relevant to real baseball.

In all sports analyses, it is important for a scientist to avoid hubris and pay careful attention to the athletes. Major league players are serious people, who are intelligent and knowledgeable about their livelihood. Specific, operational conclusions held by a consensus of players are seldom wrong, though—since baseball players are athletes, not engineers or physicists—their analyses and rationale may be imperfect. If players think they hit better after illegally drilling a hole in their bat and filling it with cork, they must be taken seriously. The reasons they give for their "improvement," however, may not be valid. I hope that nothing in the following material will be seen by a competent player of the game to be definitely contrary to his experience. If such a contradiction arises about some aspect of the game, I would presume it likely that I have either misunderstood that aspect myself or that my description of my conclusion was inadequate and therefore subject to misunderstanding.

Even as the results discussed here follow from analyses of models that can only approximate reality, the various conclusions have different degrees of reliability. Some results are quite reliable; the cork, rubber, or whatever stuffed into holes drilled in bats certainly does not increase the distance the ball can go when hit by the bat. Some results are hardly better than carefully considered guesses: How much does backspin affect the distance a long fly ball travels? Although I have tried to convey the degree of reliability of different conclusions, it may be difficult to evaluate the caveats properly. By and large, the qualitative results are usually reliable, but most of the quantitative results should be considered with some reserve, perhaps as best estimates.

In spite of their uncertainties, judiciously considered quantitative estimates are interesting and important; whatever their uncertainties, they often supplant much weaker—and sometimes erroneous—qualitative insights. Consequently, I have attempted to provide numerical values almost everywhere: sometimes when the results are somewhat uncertain, sometimes when the numbers are quite trivial but not necessarily immediately accessible to the reader.

As this exposition is directed toward those interested in baseball, not physics, I have chosen to present quantitative matters in terms of familiar units using the English system of measures—distances in feet and inches, velocities in miles per hour (mph), and forces in terms of ounce and pound weights. I have also often chosen to express effects on the velocities of batted balls in terms of deviations of the length of a ball batted 400 feet (a long home run?) under standard conditions.

To express the goals of this book, I can do no better than adopt a modification of a statement from Paul Kirkpatrick's article "Batting the Ball": *The aim of this study is not to reform baseball but to understand it.* As a corollary to the statement of purpose, I must emphasize that the book is not meant as a guide to a player; of all the ways to learn to throw and bat a ball better, an academic study of the mechanics of the actions must be least useful.

THE FLIGHT OF THE BASEBALL

THE BASEBALL—AIR RESISTANCE

From the *Official Baseball Rules:*

> 1.09 The ball should be a sphere formed by yarn wound around a small sphere of cork, rubber, or similar material covered with two stripes of white horsehide or cowhide, tightly stitched together. It shall weigh not less than 5 nor more than $5\frac{1}{4}$ ounces avoirdupois and measure no less than 9 nor more than $9\frac{1}{4}$ inches in circumference.

The description in the rule book, ingenuous and charming, is not that of an engineer; the manufacturer (once in Chicopee, Massachusetts, then Haiti, then Taiwan, and, in 1989, Haiti again) requires further directions: "The cork nucleus, enclosed in rubber, is wound with 121 yards of blue-gray wool yarn, 45 yards of white wool yarn, and 150 yards of fine cotton yarn. Core and winding are enclosed by rubber cement and a two piece cowhide [horsehide before 1974] cover hand-stitched together with just 216 raised red cotton stitches."

Much more is required to define completely the ball that is the center of the sport of baseball, but its flight is largely set by the size and weight constraints listed in the rules. The paths of base-balls projected at velocities common to the game are strongly influenced by air resistance. The forces on the ball from this resistance are typically of the same magnitude as the force of gravity. A ball batted with an initial velocity of 110 mph at an angle of 35° from the horizontal would go about 750 feet in a vacuum; at Shea Stadium in New York, it will travel only about 400 feet. Hence, it is necessary to understand the fluid dynamics of air flow about spheres to understand the flight of the baseball.

When an object (such as a baseball) passes through a fluid (such as air), the fluid affects the motion of the object as it flows about the object. Moreover, for all fluids and all objects, the character of the flow of the fluid is determined by the value of a (dimension-less) Reynolds number proportional to the density of the fluid, the fluid velocity, the size of the object, and inversely proportional to the viscosity of the fluid[a]. For a given Reynolds number, the behavior of the gaseous fluid of stars—interacting with each other through gravity—that make up a galaxy 100,000 light-years across is described in very much the same way as the behavior of the molecules of air passing through an orifice one micron across, where a micron is about equal to the resolution of a high-power microscope.

The most interesting actions in the game of baseball take place when velocities of the ball range from a few mph (and Reynolds numbers of 10,000) to values near 120 mph (and Reynolds num-bers near 200,000). For velocities in that range below about 50 mph, the flow of the air about the ball is rather smooth, though trailing vortices are generated. This air flow does not actually reach the surface of the ball where there is a quiet (Prandtl) boundary layer. A very, very small insect (perhaps a plant aphid) sitting on the moving ball would feel no breeze at all. At velocities above 200 mph the flow penetrates the boundary layer (the aphid would have to hold on very tight if it were not to be blown off), and the air at the boundary—and trailing behind the ball—is quite turbulent. We label the two regions conveniently (though a little inaccurately) as *smooth* and *turbulent.*

Thus, for a baseball passing through air at a velocity less than 50 mph the air flow is smooth, but the air flow is turbulent for velocities greater than 200 mph. Much of the subtlety of baseball is derived from the fact that so much of the game is played in the region between definitely smooth flow and definitely turbulent flow, at ball velocities greater than 50 mph and smaller than 120 mph. For balls traveling at the transition velocities between 50 mph and 120 mph, the flow can be smooth or turbulent depending on the detailed character of the surface of the ball and its motion. By and large, turbulence will be induced at lower velocities by roughness in the surface—and held off at higher velocities if the surface is very smooth. Furthermore, at a given velocity, the air resistance is, surprisingly, *smaller* for turbulent flow than for smooth flow.[1]

From our understanding of fluid flow, it is convenient to describe the force on a moving baseball[b] (or equivalent sphere) as proportional to the cross-sectional area of the ball, proportional to the square of the velocity of the ball (double the velocity, increase the drag by a factor of four), proportional to the density of the air, and proportional to a quantity called the drag coefficient, which varies slowly with the velocity of the ball. Since the size of the ball is fixed and the density of the air does not vary much for the conditions under which baseball is played, we can consider that the force on the ball is proportional to the square of the velocity of the ball and the drag coefficient, which depends only on the velocity through the value of the Reynolds number.

The solid line on the graph of Figure 2.1 shows an estimate of the variation of the drag coefficient for a baseball as a function of the velocity of the ball. The drag will also depend to some extent on the orientation of the stitches on the ball. When the ball is rotating—as is usually the case—the drag will depend on the position of the axis of rotation with respect to the body (i.e.,

---

[1]Chemical engineers build piping systems in refineries so that turbulence is induced artificially and less pressure will be required to pass the liquids through the pipes.

the stitches) of the ball, on the direction of the axis with respect to the ground and the direction of the ball's flight, and on the velocity of rotation of the ball. The drag on a rapidly spinning ball is probably slightly greater than that on a slowly rotating ball. But that effect must be small; we estimate that the incremental drag on the spinning ball will usually not be much greater than one percent of the drag on the nonspinning ball; for a 90-mph fast ball thrown with a spin of 1500 rpm, the extra drag will reduce the speed of the ball as it crosses the plate only about 0.2 mph.

For rotating balls, the dependence of the drag on these factors is not likely to be large, however, and we can consider that the drag effects described here represent a kind of average over configurations and spins. Aside from these caveats, the values of the coefficients for a baseball are not well known, but wind tunnel measurements have been made of the force on a ball at velocities up to 95 mph that support the solid curve. In particular, wind tunnel measurements showed the ball suspended nearly motion-

FIGURE 2.1 *The drag coefficient vs. ball velocity for various balls the size of a baseball. At the broken line the force on the ball equals gravity.*

less in an upward-directed 95 mph vertical airstream. Therefore, for a ball moving through the air with a velocity of 95 mph, the drag force is about equal to the weight of the ball.

The broken curved line in the figure corresponds to values of the drag coefficient that would generate a drag equal to the force of gravity. The ball suspended in the airstream is held steady by the force of the air. This is the case only if the force is greater if the ball is falling and hence moving faster through the air and if the force is weaker if the ball is rising and moving slower with respect to the air. Consequently, the observed stability of the ball demands that the solid curve cross the broken line from left to right, further defining the variation of the drag coefficient with the velocity of the baseball.

The values expressed for larger velocities up to 150 mph are estimates, albeit guided by theoretical considerations. The mean uncertainties are perhaps 10 percent for velocities less than 120 mph, which is near the highest velocity reached by a ball in play. The rather gradual reduction in the drag coefficient from the value of about 0.5 for velocities less than 50 mph to the values of about 0.2 for velocities greater than 120 mph suggests that the transition from smooth flow to turbulent flow of the air passing the baseball in flight occurs gradually.

The values of the drag coefficient for an ideally smooth ball and a uniformly rough ball—about as rough as a ball completely covered with stitches—are shown also. The variation of the drag coefficient with velocity will have the same general character for a ball that is—uniformly—a little less, or a little more, rough, but the drag minimum will be found at lower velocities for a rougher ball and at higher velocities for a less rough—or smoother—ball.[2]

At velocities of 50 to 130 mph dominant in baseball, the air passes over a smooth ball the size of a baseball in a smooth high-resistance flow. Turbulence is not induced until velocities near 200 mph. A real baseball, however, with its raised stitch

---

[2]These properties of balls of various degrees of roughness were established experimentally by E. Aschenbach and published in 1974 in the *Journal of Fluid Mechanics*.

patterns, induces low-resistance turbulent flow at the baseball velocities. Consequently, if the baseball were quite smooth rather than provided with protuberant stitches, as is the case, it could not be thrown or batted nearly as far—a stitched baseball batted 400 feet would travel only about 300 feet if it were very smooth. This effect is dramatic in golf; the air resistance from the smooth flow about a smooth ball would be so great that the ball would go nowhere. The ball is thus artificially roughened by the dimples impressed in the covering to induce turbulence and reduce the air resistance.

At velocities near 175 mph, where the resistive force for smooth balls falls off sharply, the resistance on the ball actually becomes smaller as the velocity of the ball increases. The resistance on a smooth ball the size of a baseball traveling 190 mph is smaller than the resistance on such a ball traveling 160 mph. This sharp dip in the drag coefficient at the onset of turbulence has been called the "drag crisis." For uniformly rough balls the sharp reduction in drag with increasing velocity comes at lower velocities; the rougher the ball the lower the velocity at which turbulence is induced and the lower the velocity of the drag crisis. If the baseball suffered such a drag crisis at velocities common in baseball, anomalous effects could be important.[3] For example, a ball hit against the wind might go farther than a ball hit with the wind. It seems, however, that the baseball, usually rotating in flight, presenting different configurations of smoothness and roughness to the air as a consequence of the changing orientation of the stitches, does not correspond to any *uniformly* rough ball and does not undergo the sudden transition from smooth to turbulent flow that characterizes a drag crisis.

Since the drag for a rough ball can be less than for a smoother ball, even if there is no baseball drag crisis, the distance a batted ball will travel might still depend on the character of the ball's surface. A very rough, scarred ball with a surface that could induce

---

[3]Cliff Frohlich discusses the drag on baseballs—and much else about baseball—in an article in the *American Journal of Physics* published in 1984.

FIGURE 2.2   *The solid line shows the variation with velocity of the drag force on a baseball. The broken line shows the variation with velocity of the Magnus force for a ball spinning at a rate of 1800 rpm. The forces are expressed in units of the ball's weight.*

turbulence at low velocities could well travel farther than a new smooth ball. Smaller changes could also be significant. A change in the surface of the ball was initiated in 1974, when the traditional horsehide cover was exchanged for cowhide. A judicious estimate, however, suggests that for the range of surface conditions tolerated for baseballs used in the major leagues the dependence of the drag on the character of the basic skin surface can be neglected, i.e., cows and horses are not too different. But any significant changes in the height of the stitches might change the velocity at which turbulence begins and thus affect the flight of the ball and the distance balls can be hit.

Returning to the regulation baseball, the solid curve of Figure 2.2 shows the variation of the retarding force on the ball with velocity derived from the baseball drag coefficients of Figure 2.1. The force is expressed in terms of the weight of the ball; hence, the value 1.0 of the ordinate corresponds to the force of gravity.

There may be significant differences in the drag force for different orientations of the stitch configurations with respect to the direction of motion, so the forces shown here must be considered as a kind of average over stitch orientations.

Note that the retarding force on the ball increases with velocity and is equal to the weight of the ball for a ball velocity of 95 mph. Therefore, the terminal velocity of a ball dropped from a great height is but 95 mph.

SPIN AND THE MAGNUS COEFFICIENT

The total force on a baseball, from the normal air pressure of 14.6 pounds per square inch, pushing the ball toward third base as it travels from pitcher to batter, is nearly 100 pounds. Of course, there is, ordinarily, a nearly identical force pushing the ball toward first base. If these forces differ by as much as an ounce and one-half—or about one part in a thousand—a ball thrown to the plate at a velocity of 75 mph will be deflected, or curve, a little more than a foot. Such modest, asymmetric force imbalances are generated by asymmetric spinning of the ball and by asymmetric placement of the stitches on the ball.

If the resistive force on a ball is proportional to the square of the velocity of the air passing the ball, it would seem probable that there would be such an unbalanced force on a spinning ball since the velocity through the air of one side of the ball at the spin equator is greater than the velocity of the other side. Such a force, directed at right angles to the direction of the air velocity and to the axis of spin, has long been known and is called the *Magnus* force.

Some insight into the force imbalance can be gained by considering the different forces that follow from the different velocities of the opposite sides of a spinning ball. Figure 2.3 indicates the origin of this transverse force for the "normal" situation, where the air resistance forces increase as the velocity increases. Consider that a curve ball is thrown (from left to right in the figure) by a right-handed pitcher at a speed of 70 mph so that it rotates 17 times counterclockwise (as seen from above) in its trip of about 60 feet from the pitcher's hand to the plate; such a ball will be rotating at a rate of about 1800 rpm; i.e., about one half the rate of a small synchronous electric motor. The side of the ball toward third base (at the bottom of the figure) then travels about 15 times 9″ (the circumference of the ball), equal to 11′, farther than 60′, while the side toward first base travels 11′ less than 60′. The velocity of the third-base side is then about 82 mph, and the velocity of the first-base side is only 58 mph. As shown in Figure 2.2, the drag force increases with velocity; i.e., the difference in the pressure of the air on the front face of the moving ball is greater than on the rear face, and that pressure difference increases with velocity. We can then expect the air pressure on the third-base side of the ball, which is traveling faster through the air, to be greater than the pressure on the first-base side, which is traveling more slowly, and the ball will be deflected toward first base.

If the resistive drag force varies only as the square of the velocity and if the Magnus force is only an imbalance in that resistance, which follows from the faster motion of one side of the

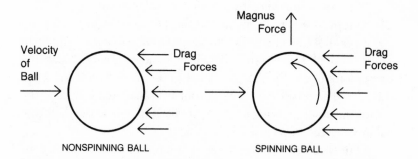

FIGURE 2.3   *The character of the drag forces on a nonspinning and spinning ball suggests the origin of the Magnus force. The imbalance in the components of forces normal to the surface of the spinning ball generates a force directed upward on the ball.*

ball through the air than the other, we should expect that the Magnus force will be proportional to spin frequency, proportional to the air velocity (or ball velocity), and proportional to the drag value of the coefficient at the ball velocity[c] V.

Though the Magnus force can be described qualitatively in terms of such an imbalance of the drag forces, a reliable quantitative description of the force on a stitched baseball is not available (more properly, measurements that should be reliable do not agree). Hence, we adopt a model of the Magnus force on a baseball that we believe must reflect, correctly, the general characteristics of the real force as expressed by the drag imbalance—and at worst—is not likely to be seriously in error. So from this view the Magnus force, described by the imbalance of resistive forces on the ball that follows from the imbalance of velocity of the air flow past the spinning ball, is proportional to the rate of change of the resistance with velocity which is just the slope of the drag resistance curve shown as the solid line in Figure 2.2. From this model[d] of the Magnus force, the magnitude of the force (expressed in units of the weight of the ball) is shown in Figure 2.2 as a broken line for balls rotating at a rate of 1800 rpm, which is near the maximum for thrown balls. We judge the uncertainties in our estimates of the magnitude of the Magnus force on the baseball plotted in Figure 2.2 as about 25 percent.

The maximum Magnus force on a ball spinning at a rate of 1800 rpm is seen to be about one third of the weight of the ball, so we cannot expect a ball spinning at that rate to curve more than one third of the distance it will fall under gravity. Since the variation of velocity is proportional to the rate of spin, the Magnus force on a ball will be proportional to that spin rate, e.g., the force on a ball spinning at a rate of 900 rpm will be one half that shown by the broken line in the figure.

Note that the Magnus force on a ball spinning at a constant rate increases with velocity up to speeds near 60 mph—the speed of a curve ball—and then probably falls off considerably. According to this model of the force, for a given spin rate the Magnus force is smaller for a ball thrown at a velocity of 90 mph—such as a major league fast ball—than for a 60-mph ball. We might expect then that a hard-thrown ball will curve much less than a slower pitch, though the spin rates might be the same. The Magnus force can be expected to be less on the fast ball, and the force has less time in which to act as the ball reaches the plate quicker.

The forces that cause the ball to be deflected must also generate a torque that slows down the spin[e]. For a hard-hit ball traveling with an initial velocity of 110 mph, we use the model of the Magnus force we have adopted to estimate that the spin rate will decrease at a rate of about 30 percent per second. For a 400-foot home run, the backspin applied by the bat (perhaps 2000 rpm) is reduced to about 350 rpm when the ball lands about 5 seconds later. This is consistent with experience; fly balls do not spin out of the gloves of outfielders.

DISTANCE OF FLIGHT OF A BATTED BALL

Projected by arm or bat, the distance a ball will travel can be calculated using the simple ballistic relations governing a body in flight and taking the retarding drag force in the direction of motion and the Magnus force, normal to the direction of motion, from the values shown in Figure 2.2. Typical trajectories are shown in Figure 2.4, and a graph of the maximum distance versus the initial ball velocity is shown in Figure 2.5. From these calculations, the maximum distance is obtained with balls projected at an initial angle of about 35° from the horizontal, though balls projected at 30° or 40° travel almost as far. Note that the ball falls at a rather large angle at the end of its flight; the trajectories are not symmetric. We find that the 385-foot fly ball hit at the optimum angle of about 35° will be in the air for about 5 seconds; a really high fly ball—or pop fly—will stay in the air more than

FIGURE 2.4 *The trajectories of balls projected at an angle of 35° with different velocities. The balls are assumed to be rotating with an initial backspin of 2000 rpm. The solid triangles show positions of the ball at intervals of one-fourth of a second beginning at a time one second after the ball leaves the bat.*

6 seconds. To put this into perspective, an average right-handed batter will run from home to first in about 4.3 seconds.[4]

Obviously these trajectories, and maximum distances, depend on the drag and the drag coefficients, which are imperfectly known. Figure 2.6 shows the variation of maximum distance the ball travels as the average drag coefficient varies. Using this graph, we conclude that the uncertainty of 10 percent in the drag coef-

---

[4]A very fast left-handed batter will reach first base in about 3.7 seconds after a drag bunt, a few tenths of a second—a step—faster than an equally fast right-handed batter, while an unusually slow runner, such as catcher Ernie Lombardi of the 1940s, might take more than 5 seconds to go from home to first. In a timed racelike effort in 1921, Maurice Archdeacon circled the bases in 13.4 seconds; the legendary James "Cool Papa" Bell of the Homestead Grays is said to have made it in less than 13 seconds. To put these times into context, a world-class sprinter like Carl Lewis, running with track shoes on a carefully prepared track, would take about 11 seconds to run the 120 yards if the bases were laid out in a line and about 11.6 seconds to run in a straight line the 127 yards that a reasonable course about the bases would measure.

ficient at high velocities translates into an uncertainty in the maximum distances of flight of about 3.5 percent or an uncertainty of 14 feet for a 400-foot home run. Since we are primarily interested in *differences* in distance that follow from changes in conditions, such an accuracy is acceptable.

As a ball is thrown or batted for distance in a real situation, it usually has appreciable backspin (of the order of 2000 revolutions per minute), which generates a significant force perpendicular to the direction of motion and generally upward. Though the spin will probably increase the drag on the ball slightly, the Magnus force will somewhat increase the distance traveled by balls hit or thrown at smaller angles and reduce the distance of balls hit or thrown at larger angles. In general, this effect will produce a modest increase in the maximum distances shown in Figure 2.5 and will reduce the optimum angle of projection a few degrees. A ball launched without backspin at an angle of 35° that would

FIGURE 2.5 *The distance a baseball travels as a function of the initial velocity of the ball when projected at an angle of 35° with an initial backspin of 2000 rpm. The solid circle and triangle show the distances as affected by a 10 mph wind blowing in and out.*

travel 380 feet would travel about 400 feet if the backspin were about 2000 rpm. (Coming off the bat, the ball would rotate about once in every 5 feet of flight.) Since the magnitude of the backspin depends in detail on the exact manner in which the ball is thrown or struck by the bat, I have assumed a mean backspin of 2000 rpm for simplicity of exposition in calculating the representative trajectories and distances given in Figures 2.4 and 2.5.

Obviously, the distance a ball will travel is strongly affected by the wind. Figure 2.5 also shows the effect of winds of 10 miles per hour (near the average wind velocity throughout the United States) on the flight of a 400-foot home run. With the wind behind the batter, the ball will go about 30 feet farther; with the wind against the batter, the 400-foot home run to center field will end up as a 370-foot fly ball out. In general, ball parks are laid out so that the line from "home base through the pitcher's plate to second base shall run East-Northeast"[5] so the batter will not face Nolan Ryan's fast ball in the late afternoon with the sun in his eyes. Hence, the prevailing westerlies of the Northern Hemisphere tend to blow out toward right-center field, helping the batter. Of course, for most ball parks, the effect of the wind is reduced because of the protection the stands afford.

Since the retarding force on the ball depends on the density of the air, the ball will travel farther in ball parks at high altitudes. A 400-foot home run at Yankee Stadium, near sea level, would translate to a 408-foot drive at Atlanta, which, at an altitude of 1,050 feet, is the highest city in the majors. The ball could be expected to go about 6 feet farther in Kansas City and 5 feet farther at County Stadium in Milwaukee or Wrigley Field in Chicago than at Yankee Stadium. These differences are not so great as to modify the game, but if a major league franchise is someday granted to Denver (altitude 5,280 feet) consideration should be given to requiring a larger park or using a less lively ball (special, less lively, high-altitude balls are used in tennis, though

---

[5] *Official Baseball Rules,* Section 1.04.

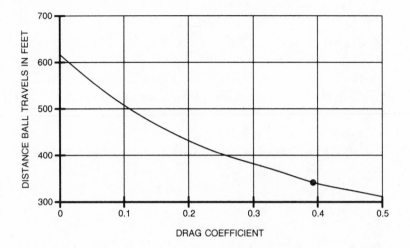

FIGURE 2.6 *The dependence of the distance a ball is hit on the drag coefficient, assuming that the coefficient was independent of velocity. The solid circle shows the distance calculated if the coefficient varies as shown in Figure 2.1.*

for somewhat different reasons). A 400-foot drive by Dave Winfield in New York would go as much as 40 feet farther through the thinner air in Denver! Furthermore, pitchers would find that their fast balls were half a foot quicker in Denver though their curves would not bite quite so well.

Players for the American Association Denver Bears generally hit a lot of home runs at home and have commented on problems playing the outfield; "fly balls come at you faster and sail farther than you might expect."

Temperature, barometric pressure, and humidity also affect the ball's flight. The standard 400-foot home run will go about 6 feet farther for a one-inch reduction in the barometer and as much as 20 feet farther on a hot 95° summer day in Atlanta than a cold 45° spring day in Milwaukee[e]. Likewise, the effect of temperature differences on the elasticity of the ball also has an effect on the distance a batted ball travels.

The humidity, per se, has little effect on the ball's flight. Indeed, since water vapor is a little lighter than air, if all other factors are the same, a ball will travel slightly *farther* if the humidity is high. The humidity, however, effects the weight and elasticity of balls in storage. Balls stored under conditions of high humidity will gain some weight through the absorption of water from the air and their elasticity (the coefficient of restitution, discussed in Chapter 4) will be reduced[g].

Since the retarding force is proportional to the cross-sectional area of the ball, small balls will go farther than large balls. According to the *Official Baseball Rules*, a ball may be as large as $9\frac{1}{4}$ inches in circumference and as small as 9 inches, as light as 5 ounces and as heavy as $5\frac{1}{4}$ ounces. The stroke that propels the larger ball 400 feet will drive the smaller ball—of the same weight and elasticity—perhaps 6 feet farther. The stroke that drives the lighter ball 400 feet will drive the heavier ball—of the same size and elasticity—the same distance, give or take a foot or two. Though the heavier ball will come off the bat a little slower, its greater sectional density will carry it better through the drag of the air and just about compensate for the deficit in initial velocity.

TECHNICAL NOTES

a. The value of the Reynolds number R for a sphere of diameter D moving with a velocity V through a fluid of density ρ and viscosity η is $R = \dfrac{\rho VD}{\eta}$.

b. The relation for the drag force F_d can be written as

$$F_d = C_d \rho A v^2.$$

Here $A = \pi r^2$, with r the radius of the sphere, is the cross-sectional area of the sphere, ρ is the density of the air, v is the velocity of the ball, and C_d is the drag coefficient. For $C_d = \frac{1}{2}$, this is just the force required to move the column of air displaced by the motion of the ball to match the velocity of the ball.

c. For velocities such that the drag coefficient, C_d does not vary strongly with the velocity V of the ball through the air, the Magnus force can be expressed as

$$F_m = K\omega V C_v$$

where if the Magnus force F_m is measured in *lbs* force, the ball velocity V is expressed in mph, and the spin frequency ω is measured in rpm, K has the value $2 \cdot 10^{-6}$. That value of K is taken from the results of measurements by Briggs; if the drag coefficient C_V does not vary with velocity, K is proportional to C_V. Briggs concluded from his measurements that the Magnus force varied as $V^2\omega$ rather than $V\omega$. His measurements, however, were not conclusive (he may have seen a non-Magnus effect like a knuckle ball at low spin values, leading him to erroneous conclusions), and a Magnus force proportional to V^2 is difficult to understand. A Magnus force proportional to ωV is suggested by standard fluid dynamics.

Incidentally, the Magnus effect responsible for the curve of the curve ball and the hop of the fast ball is not the same as the Bernoulli effect—which is why it is called the Magnus effect and not the Bernoulli effect. But the classical Bernoulli effect may have a role in the movement of the knuckle ball.

d. In general, the drag coefficient C_V varies strongly with the velocity (as is the case for velocities near the transition between smooth and turbulent flow), and we must consider variations in the drag coefficient C_V. We have then:

$$F_m = K\omega V C_V [1 + 0.5 \times (V/C_V) \times (dC_V/dV)]$$

According to this relation, the Magnus force can reverse sign when the logarithmic derivative $(V/C_V)(dC_V/dV)$ is less than -2; the drag force then *decreases* as the velocity increases.

Wind tunnel measurements of the Magnus coefficient have led to varying—and inconsistent—results. In particular, Briggs reported anti-Magnus effects for smooth balls at rather low velocities. My simple experiments observing the curvature of the trajectories of table-tennis balls thrown softly with "baseball" spins always gave positive Magnus effects, i.e., normal curves. For smooth balls, near the drag crisis velocity, the drag actually decreases as the velocity increases. In that case, according to the formula above, the sign of the Magnus effect will be reversed. Briggs did see such reversals in the course of measurements of the Magnus force on smooth balls.

e. We estimate the torque L, which resists the spin, as

$$L = kF_m r$$

where F_m is the Magnus force, r is the radius of the ball, and k is a proportionality constant with a value that we estimate as about equal to $\frac{1}{5}$. The energy-per-second P lost to the spin kinetic energy T is then:

$$P = L\omega = kF_m r\omega$$

where ω is the angular spin velocity of the ball. The spin-down time constant τ is then:

$$\tau = T/P \text{ where } T = \tfrac{1}{5}mr^2\omega^2$$

where m is the mass of the ball. The time constant τ is a measure of the time required for the ball to lose its spin; for example, if $\tau = 10$ seconds, the spin rate will decrease by a factor very near $1/\tau = \frac{1}{10}$ in one second.

f. The drag resistance is proportional to the density of the air and hence, for a given temperature, to the barometric pressure; a reduction in pressure of $\frac{1}{3}''$ of mercury on the barometer decreases the drag by about 1 percent. On the average, the pressure—and the drag—decreases by about 1 percent per 275-foot increase in altitude. The density of the air is also reduced by about 1 percent for each 5°F rise in temperature, but the mean drag may decrease a little more than 1 percent since the viscosity of the air will increase about $\frac{1}{2}$ percent (and the Reynolds number will decrease by about $\frac{1}{2}$ percent). For the standard 400-foot home run, a decrease in the drag by 1 percent will add about 2 feet to the ball's carry.

g. R. C. Larsen found that the weight of balls stored at 100 percent humidity for four weeks increased by 11 percent and the coefficient of

restitution at an impact velocity of 25 mph decreased by 10 percent—
when dropped on concrete from a height of 20 feet, the humidified balls
will bounce only about 80 percent as high as the balls stored at low
humidity. If that proportional decrease in elasticity would hold at larger
impact velocities, the swing of the bat that would drive a "dry" ball 380
feet would propel the ball stored at high humidity only 350 feet. After
sitting in dry air for a few hours to dry the cover, the humidified balls
could not be distinguished from normal balls by a layman. An experi-
enced pitcher, however, would probably notice that they were heavier
and softer than the balls he was used to.

PITCHING

THE PITCHER'S TASK

The most important person on the team, in any one game, is the pitcher.[1] The pitcher's job is simply defined—if not simply executed. From the pitcher's mound, 60′6″ from the rear point of home plate, he must throw the ball over the plate for strikes. But within that constraint he must project the ball in patterns of trajectories, velocities, and placements so that the batter cannot hit the ball squarely. But what is a strike?

A STRIKE is a legal pitch when so called by the umpire, which (b) is not struck at, if any part of the ball pass through any part of the strike zone.

The STRIKE ZONE is that space over home plate which is between the batter's armpits and the top of his knees when he assumes his natural stance [*Official Baseball Rules*, 1987, page 22].

1.05 Home base . . . [is] 17 inches long . . . with the 17-inch edge facing the pitcher's plate [*Official Baseball Rules*, page 5].

---

[1]Though over the whole season, the best hitter in baseball—Babe Ruth—was judged to be more valuable than one of the best pitchers in baseball—Babe Ruth. In 1919, Ruth was moved from the Boston Red Sox pitcher's box, where he worked every fourth day, to an everyday job in right field.

A typical six-foot-tall batter stands naturally at the plate with his armpits about 46" above the plate. The tops of his knees will be about 22" above the plate. With these definitions, a ball he does not swing at will be called a strike if it passes over the plate so that it knicks a five-sided box, with a cross section the shape of the plate, 17" wide, 24" high, and 17" deep, which might be suspended with the bottom edge 22" above the plate. As a matter of practice, umpires define the high boundary of the strike zone about 4" lower than the armpits so the effective zone is about 17" by 20". Hence, the center of the ball must strike a three-dimensional target about 20" wide by 23" high and 17" deep. Control pitchers hit corners with an uncertainty of about 3". One must be a fairly good shot to shoot a pistol with that accuracy. A mistake, usually an error of a foot that puts the ball in the center of the strike zone, on one pitch of a hundred, may lose the ball in the bleachers—and the game.

The pitcher must not only throw the ball so that it passes through the strike zone, he must throw so that the batter does not hit the ball squarely. He does this by throwing the ball so that it passes through specific places at or near the strike zone that the batter finds difficult to reach, by throwing the ball at different velocities to upset the batter's timing, and by applying spin to the ball in such a manner that the ball passes the batter with different trajectories, confusing and confuting him. Thus, the pitcher varies the placement, the velocity, and the movement of the ball. The pitcher's action up to the release of the ball is part of the art of pitching; the action of the ball after release, determined by the laws of nature, is addressed by physics and is subject to our analyses.

THE CURVE BALL

Interest in the left-right curvature of balls sailing through the air is probably as old as ball games themselves. Isaac Newton, at the age of 23, discussed the curvature of tennis balls in terms that make good sense today. In the nineteenth-century genesis of mathematical physics, Lord Rayleigh analyzed the curvature of the path of spinning balls, and P. G. Tait, the eminent Scottish physicist, wrote extensively on the curves of golf balls—perhaps in an attempt to understand and cure a slice. The Baseball Hall of Fame in Cooperstown, New York, gives precedence to Candy Cummings as the first baseball pitcher to throw a curve ball, though the rules of the 1860s required Cummings to throw underhand rather as a softball pitcher today. Freddy Goldsmith and others confounded batters with curve balls at about the same time.

Balls curve as a consequence of asymmetries in the resistance of the air through which they pass. If the air resistance is greater on the third-base side than on the first-base side of a ball thrown from the pitcher to the batter, the ball will be forced—curve— toward first. Aside from generating curved paths, this resistance affects the flight of the pitched ball by reducing the velocity of a ball thrown from the pitcher's mound at a speed of 100 mph to a velocity of 90 mph as it crosses the plate about 0.40 seconds after it leaves the pitcher's hand. If the ball is not spinning very fast, during the time of flight it will fall almost three feet below the original flight line. If the ball is rotating quickly, differences in the force of the air on the ball transverse to the spin axis are induced that cause the trajectory of the ball to deviate from the original horizontal direction of motion and deviate vertically from the trajectory determined solely by gravity. Such asymmetric forces follow from the rotation of the ball when a curve or hopping (rising) fast ball is thrown or from differential forces on the seams for a slowly rotating knuckle ball. The action of a curve ball and knuckle ball are quite different.

First we consider the curve ball.

As we have noted, the force on a baseball is greater on the side of the ball that passes more quickly through the air due to the Magnus effect. The diagram at the top of Figure 3.1 shows the trajectory (reconstructed from Briggs's wind tunnel measurements) of a rather wide-breaking curve ball thrown so as to rotate counterclockwise—as seen from above in the figure—by a right-handed pitcher. This ball is thrown with an initial velocity of 70 mph, spinning at a rate of 1600 rpm, to cross the plate about 0.6 seconds later at a speed of about 61 mph. Although the radius of curvature is nearly constant throughout the ball's flight, the deflection from the original direction increases approximately quadratically with distance, i.e., four times the deflection at twice the distance. Halfway from pitcher to the plate, the ball has moved about 3.4 inches from the original line of flight, which is directed toward the inside corner and is moving toward the center of the plate. At the plate, the deflection is 14.4 inches and the ball passes over the outside corner. From the perspective of the batter—or pitcher—the ball that started toward the inside corner has "curved" 14.4 inches to pass over the outside corner. Moreover, one half of the deflection occurred during the last 15 feet of the path to the plate. (We realize that the most useful curve from a tactical view curves down much more than sideways, but we discuss the transverse motion for expositional simplicity.) Does a curve ball then travel in a smooth arc like the arc of a circle? Yes. Does the ball "break" as it nears the plate? Yes. Neither the

FIGURE 3.1   *The trajectories of a curve ball and a knuckle ball on their way from a right-handed pitcher to a batter. The curve ball is rotating counterclockwise as viewed from above the ball's line of flight.*

smooth arc nor the break is an illusion but a different description of the same reality.

Though the deflection as seen, correctly, by batter and pitcher is 14.4″, the sagitta—the largest deviation from the straight line drawn from the beginning to the end of the ball's flight as shown in the figure—is but 3.4″. Hence, it is difficult to throw a ball with a diameter of 2.9″ through three aligned rods so that the ball will pass to the left of one rod, to the right of the second, and to the left of the third. In the course of arguments in 1870 as to whether a curve ball really curves, Freddy Goldsmith performed that feat in New Haven.

But Goldsmith probably threw a slower curve. For slow curves thrown with a definite spin rate, the deflection is approximately proportional to the time the ball is in the air. Hence, a ball thrown with an initial velocity of 65 mph with a 1600 rpm spin, which takes about 8 percent longer to reach the plate than the 70 mph pitch, will curve about 8 percent more.

Conversely, according to the model of the Magnus force we have adopted, it is almost impossible to throw a fast ball that curves strongly. As shown in Figure 2.2, the transverse Magnus force that induces the curved trajectory is smaller for velocities greater than 70 mph. Also, since the faster ball reaches the plate sooner, the force has a shorter time in which to act; for the same

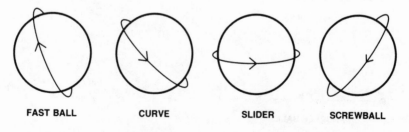

FAST BALL   CURVE   SLIDER   SCREWBALL

FIGURE 3.2 *Ball rotation directions, as seen by the batter, for pitches thrown almost straight overhand by a right-handed pitcher. The arrow shows the direction of rotation, which is also the direction of the Magnus force.*

transverse force, the ball that travels 10 percent faster will curve 20 percent less.

There are other pitches of interest. The slider—sometimes called a "nickel curve" 70 years ago—is a kind of fast curve. Thrown at a higher velocity than the standard curve ball, the break of the slider is smaller than the deflection of the curve ball and the spin axis is such that the deflection is more nearly left-right than the curve—which, at best, is more of a pure drop. The screwball thrown by Carl Hubbell and others, and called the fadeaway by Christy Mathewson, is a kind of reverse curve thrown by a right-handed pitcher to break away from a left-handed batter a little like a left-handed curve ball.

Figure 3.2 shows typical spin directions for different pitches.[2]

THE KNUCKLE BALL

The thrown ball can also be deflected by the turbulence induced by the stitching on the flow of air passing by the ball. If the ball is thrown with very little rotation, asymmetric stitch configurations can be generated that lead to large imbalances of forces and extraordinary excursions in trajectory. Low-resistance turbulent flow can be induced by stitches on one side of the ball while the air will flow smoothly—and with larger resistance—past a smooth face on the other side. We can be more specific. Noting that the drag on a ball is proportional to the drag coefficient, we see from Figure 2.1 that the drag on a smooth ball thrown at a velocity of 65 mph is far greater than the drag on a rough ball. We might then expect that a ball that is smooth on one side and rough on the other will encounter an asymmetric drag force—larger on the

---

[2]It is interesting to observe the curvature in the flight of a table-tennis ball produced by throwing the ball so that it spins in the directions shown in Figure 3.2. Weighing one-sixtieth as much as a baseball, very small Magnus forces will generate substantial curves in the trajectories of the light ball over a span of 20 feet.

smooth side—that will tend to deflect the ball toward the rougher, stitched side.

Measurements[3] of these asymmetric forces have been made on balls mounted in wind tunnels that allow the calculations of trajectories of a pitched ball. The diagram at the bottom of Figure 3.1 shows such a trajectory (albeit an exceptionally dramatic case). Thrown originally toward the center of the plate, when the ball is only 20 feet from the plate it is 11 inches off-center and heading toward the visiting team's dugout. The catcher starts moving desperately to his right to avoid a wild pitch, and the right-handed batter relaxes, knowing the ball will be two feet outside. Then, confounding catcher, batter (and pitcher), the ball ducks over the center of the plate for a called strike—and passed ball.

Other examples of a fluttering ball—which curves in both directions!—were also generated by the wind tunnel simulation. In general, the knuckle-ball pitcher tries to throw the ball with a small rotation (about one-half rotation from pitcher to batter) so that the stitch configurations—and forces on the ball—change on the way to the plate. Such a slowly rotating ball can be thrown off the knuckles—with the ball held between the forefinger and little finger—but the knuckle ball in practice is usually thrown off the fingertips or fingernails—which are carefully cut square.

The disadvantage of the knuckle ball from the view of pitcher, catcher, and manager is that the forces can vary strongly with very small differences in ball orientation. Hence, the pitcher, however skilled, finds it very difficult to control the pitch.

THE HOP OF THE FAST BALL

The fast ball may be the key baseball pitch; surely it is the most dramatic. How fast can a pitcher throw a ball? It seems that the

---

[3]These measurements were described by R. G. Watts and E. Sawyer in the *American Journal of Physics* in 1975.

fastest pitchers can throw the ball so that it crosses the plate with a velocity of about 100 mph. In 1946, Bob Feller threw a ball that was measured to have a velocity of 98.6 mph as it crossed the plate. In 1914, a pitch of Walter Johnson was found to have a velocity of 99.7 mph. Nolan Ryan threw a ball with a velocity timed at 100.7 mph. Pitches by J. R. Richards, Goose Gossage, and Lee Smith have also been determined to cross the plate with velocities near 100 mph. Among the old-timers, Lefty Grove and Smokey Joe Wood probably threw with comparable velocities according to contemporary judgments, and Satchel Paige and Slim Jones threw bullets in the Negro Leagues of the 1930s. Grove—and some batters—thought that he was faster than Feller, and Walter Johnson once said that "no one ever threw harder than Smokey Joe Wood." (Dr. Joe Wood after receiving an honorary degree from Yale in the 1970s.) Though adequate velocity measurements are possible with instruments as simple as the ballistic pendulum used in 1914 and as sophisticated as the radar guns used today, some care is required to get an accurate result with either technique.

Since the ball slows down considerably on the way from the pitcher to the plate, the "muzzle velocity" of the ball—as it leaves the pitcher's hand—is about 8 mph greater than its speed across the plate. The ball loses speed at the rate of about 1 mph every 7 feet.

It is well known that the backspin applied to the overhand fast ball causes the ball to rise—or hop; such a ball will be thrown with a backspin of perhaps 1200 rpm and rotate about eight times on its way from pitcher to plate. Though the hop is not likely to be much greater than 4 inches, this is more than enough to trouble the batter swinging a 2.75-inch-diameter bat because he must initiate his swing when the ball is about halfway to the plate and the deviation from the hop is only about 1 inch.

Like its cousin, the curve ball, the hopping fast ball follows a smooth arc on its way to the plate, but half of the hop deviation occurs in the last 15 feet of that flight.

Figure 3.3 shows the trajectories of a typical 90-mph fast ball

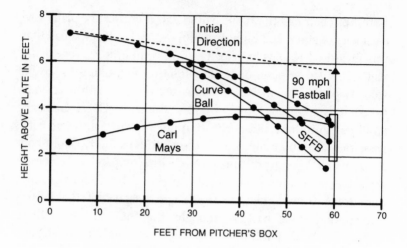

FIGURE 3.3   *Elevation views of the trajectories of pitches from the pitcher to the batter. Ball velocities are measured in mph, rotations in rpm. The points show the positions of the ball at intervals of one twentieth of a second. The different horizontal and vertical scales distort the trajectories.*

FIGURE 3.4   *Trajectories—with no scale distortion—of different pitches as the balls cross the plate. The 90 mph and 95 mph fast balls were launched with the same initial trajectory. The ball images show positions at times that differ by one fortieth of a second.*

thrown "high and tight" together with trajectories of a curve ball at the knees and a split-finger fast ball (SFFB), all thrown from the pitcher's hand in the direction marked by the straight line. The trajectory of an underhand pitch, such as the blazing fast ball Carl Mays threw from his shoe tops in the 1920s, is also shown to demonstrate the variety of slants the batter must face. Figure 3.4 shows the trajectories, undistorted by the scale differences, of certain pitches as they cross the plate.

The relatively small hop of 4 inches or 5 inches may seem somewhat surprising. But the "upward curve" of 5 inches is not very different from the "tailing off" seen in fast balls thrown sidearm (or from a three-quarter overhand pitch), in which the axis of rotation is tilted so the hop has a left-right component that is more clearly seen. Again, the small break follows from the short time the force acts and from the smaller Magnus coefficient at higher velocities—as shown in Figure 2.2. A 75-mph (high school) fast ball, thrown with the same motion—although with less intensity—may actually hop more since the Magnus force (as shown in Figure 2.2) may be larger and the force would be applied for a longer time to the slower ball.

The magnitude of the Magnus forces and drag forces presented in Figure 2.2 represent a kind of average over different stitch configurations. Since the pitcher can control those configurations, he can produce somewhat different effects by releasing the ball so that it spins with the axis oriented differently with respect to the stitch pattern. The diagram at the left of Figure 3.5 shows a ball held for the delivery of a "with-the-seams" fast ball, in which two seams pass the equator at every revolution. At the right, the release position for the "cross-seam" fast ball is shown and four seams pass the equator per revolution.

It is plausible that the aerodynamics of the ball delivered with such different spin axes will be significantly different. An inspection of the graph of Figure 2.1 illustrating the variation of the drag coefficients with velocity for balls of different smoothness shows that the resistance on a 95-mph fast ball would be about two-thirds greater if the ball were very smooth. It is certainly

FIGURE 3.5 *At the left is a conventional with-the-seams fast ball release configuration with a scuffed area situated asymmetrically. At the right, the release configuration of the cross-seam fast ball is shown.*

plausible that the with-the-seam rotation, showing only two seams per revolution, has a smoother than average profile in the airstream and that the cross-seams rotation might appear rougher than average. If this is the case, we could expect a greater-than-average drag on the with-the-seam fast ball and, hence, a somewhat smaller than average drag on the cross-seam delivery.

How large can we expect such effects to be? Using the "average" drag resistance we estimate that a pitcher must deliver the ball with a muzzle velocity near 100 mph if the ball is to cross the plate with a velocity of 90 mph (an average speed of 95 mph). If the drag were reduced by half for the cross-seam delivery, the ball would cross the plate with a velocity of 95 mph (and an average velocity of 97.5 mph) and reach the batter almost a foot and a half faster. If the drag were increased by two-thirds, as for a *perfectly smooth ball,* the ball thrown with an initial velocity of 100 mph would be slowed down to only 83 mph at the plate (for an average velocity of 91.5 mph) and reach the plate about 2 feet behind the "standard" pitch. But these are extremes, holding for an ideally smooth ball and a particular uniformly rough ball. Real baseballs are far from either. It is plausible, however, that differences in cross-the-plate velocities as great as 4 mph might occur because of the different spin axes that result from different ball grips. That would translate to a different average velocity of about 2 mph, or a little over a foot, on the fast ball.

Since the curve, or hop, on a ball follows from a difference of the drag on the two sides of the ball, we might expect that the spin axis that generated less drag would also generate less curve, or hop. Then, if the cross-seam fast ball moves faster, it might hop less.

Some pitchers purposefully throw a "sinking" fast ball, with little spin and that about an axis that is relatively ineffective in generating a Magnus force. Such a pitcher might hold the ball as shown in the left-hand drawing of Figure 3.5, except with his fingers set on the smooth leather rather than the stitches so the ball will slide from his hand with little spin. This ball, delivered without much backspin, will undergo little upward Magnus force and come to the plate lower by as much as 4 to 6 inches than the hopping fast ball. Delivered at the knees, the batter will tend to hit the top of the ball and roll out weakly to the infielders.

There are long-standing arguments about the magnitude of the rise of an overhand fast ball. Can such a pitch, thrown with a lot of backspin, actually rise as it heads toward the plate? From the trajectories shown in Figure 3.4, the answer would seem to be no. But, as we have pointed out, we do not understand the ball's flight through the air as well as we would like. Could our arguments be wrong, and could the players who swear that some pitchers throw a rising ball be right?

We can cast some light on the problem by noting that since the 90-mph fast ball must fall almost 3 feet on its way to the plate if there is no hop, the effect of the backspin must be such as to "curve" the ball upward about 3 feet if the ball is to rise.[4] But the magnitude of that hop will be duplicated by a pitcher who throws a 90-mph fast ball sidearm, with a sidespin instead of a backspin, only that curve will be from left to right (for a right-

---

[4]Indeed, since over-the-top pitchers must throw the ball slightly down to deliver a strike at the belt, the ball must curve up as much as 3 feet more than the 3-foot gravity effect if it is to actually rise as it crosses the plate in the strike zone.

handed pitcher). But who has seen a fast ball curve 3 feet? In fact, a hard fast ball thrown sidearm (or with a three-quarter motion) will tail off no more than 6 inches, if that much. Therefore, overhand fast balls certainly do not rise, and our calculations are not likely to be wildly wrong.

Good softball pitchers, throwing underhand upward to a batter 46 feet away from a pitcher's box that is not raised on a 10-inch mound, *do* throw balls that rise as they pass the batter's armpits. However, the underhand baseball pitchers such as Carl Mays in the 1920s and Dan Quisenberry and Kent Tekulve today, who throw 60'6" from an elevation of 10 inches, can't really get the ball to rise as it passes the batter.

If the baseball fast ball falls nearly 3 feet on its way from pitcher to plate, how can thoughtful, intelligent players possibly believe it might rise? Here we have a matter of perception. If one draws a straight line from the pitcher's hand to the position of the ball as it crosses the plate as shown in Figure 3.3, the peak of the ball's actual trajectory will lie only about 8 inches above that line. Over the distance of about 56 feet from hand to plate, the real path then differs only about 8 inches from a straight line. The 3-foot drop is from the line projected from the ball's flight as it leaves the pitcher's hand. (This is the line that would describe the ball's flight if the game were played in outer space, where there is no gravity or air.) The player who sees the ordinary fast ball as traveling almost in a straight line is not wrong. We can see that he might reasonably consider that a ball that arrives above the usual line by 6 inches must rise.

Similar misunderstandings occur about the flight of a throw from third to first by a strong-armed third baseman. Some players state emphatically that a hard-thrown ball travels in a straight line, but a physicist concludes that the ball must drop about 10 feet. Again, the 10 feet is the drop below the initial line of projection. The peak trajectory of the ball is only about $2\frac{1}{2}$ feet above a straight line drawn from the third baseman's hand to the first baseman's mitt. And a $2\frac{1}{2}$-foot deviation over the 127-foot distance between third and first is not far from a straight line.

Moreover, since fields are landscaped with the center of the diamond higher than the base baths, so the infield will drain toward the foul lines, and the pitcher's mound is 10 inches high, the ball thrown and caught at a height of 6 feet can pass near the ear of the six-foot-tall pitcher. And the pitcher may well have to duck to avoid the catcher's throw to the shoe tops of the second baseman in position to tag the runner attempting to steal second.

The ordinary overhand fast ball is thrown so that the ball rolls off the first and middle fingers of the pitching hand—indeed, these fingers usually rest on the stitches to accentuate the grip and the spin. If the pitcher places the ball so that he holds its axis between his two widely split fingers at about the first joint, he will propel the ball almost as fast but with very little spin. The curves of Figures 3.3 and 3.4 show trajectories of such well-thrown split-finger fast balls.[5] Usually the split-finger fast ball is not projected quite so efficiently as the regular backspinning fast ball and does not achieve quite the same velocity. With the same initial trajectory as the hopping 90-mph fast ball directed across the letters, the 80-mph split-finger fast ball reaches the plate about 16 inches lower and at the knees—and about 7 feet behind the fast ball.[6]

PITCHING WITH AN ILLEGALLY MODIFIED BALL

From baseball's beginnings, pitchers have enhanced their skills at throwing the ball with different velocities and spins by modifying the ball and the pitching hand. Saliva, sweat from the forehead, or surreptitious Vaseline applied to the pitching hand aid in the delivery of the slowly rotating "spitball" by allowing the ball to slip from the hand without spin. The main effect of the lubricant

---

[5] If the ball is jammed further toward the hand, the velocity of the pitched ball is reduced considerably and the pitch is more of a pure change-of-pace. Such a pitch is often called a "fork ball." However, in practice, the term is used variously.

[6] If they were released by two pitchers at the same time, the split-finger fast ball would lag behind the fast ball by 7 feet as they crossed the plate.

is that it allows less skilled pitchers to emulate what a skilled knuckle ball pitcher can accomplish legally—though the spitball can be thrown harder than the knuckle ball. The application of a lubricant to the forefinger and middle finger on delivery of an overhand fast ball can produce some of the same effects as the split-finger fast ball: The ball will slide off the fingers, accumulating less backspin and then drop more than the usual fast ball thrown with backspin.

Scarring or scuffing the ball can produce asymmetric forces on the ball that result in aberrant trajectories. Since it is probably impractical to scuff or scrape the surface of the ball so that the imperfection has as dominant an effect as the stitching, it might seem that such modifications cannot be very important, and that is largely the case for casual throws. But the highly skilled pitcher[7] can throw the ball so that the effect of the stitches is symmetric, but if the ball is scuffed on one axis—and not the other—unbalanced forces can be realized, which act only in one direction, as suggested in the left-hand diagram of Figure 3.5. Properly thrown, the scuffed left-hand side of the ball could induce low-resistance turbulence in the air passing by while the air will pass the smooth right-hand side in high-resistance smooth flow and the ball will veer to the left—*toward* the scuffed area! Since a deviation of a fraction of an inch can change a home run into a pop fly or a double-play ground ball, the controlled deviations a skilled pitcher can induce by disfiguring the ball are important.

The curve—called "swing"—of the bowled cricket ball plays a part in that game analogous to the curve in baseball pitching. Throwing on the run, the cricket bowler, projecting the ball with an overhand straight-arm motion, sends it to the batter on the bounce—and almost as fast as a fast ball from a baseball pitcher. The cricket ball is about the same size and weight as a baseball,

---

[7]This scuff-ball effect can also be simulated by a table-tennis ball thrown indoors. To conduct this test, take a square of Scotch tape—perhaps $\frac{3}{8}''$ × $\frac{3}{8}''$—and stick it on the ball so it wrinkles. Then throw the ball with backspin as suggested by the left-hand diagram of Figure 3.5 with the tape on one, or the other, pole of the axis of rotation. The ball will veer, counterintuitively, *toward* the tape[a].

but the raised stitches form a circle about the ball's equator. Moreover, the bowler is allowed to work on the cover of the ball usually shining one side and roughening the other. The bowler, therefore, does not rely on the Magnus effect but on a kind of scuff-ball effect to project a ball that evades the batter and knocks down the wicket.

THE ENERGETICS OF A THROWN BALL

A player's actions in throwing a ball hard are complex, and the analysis of the throw lies more in the realm of physiology than physics, but the results of a calculation of the projected ball's total energy and an estimate of the rate of energy transferred to the ball, i.e., the power, in throwing provide some illumination of the character of the process.

In this estimate, we consider a pitcher throwing a major league fast ball so that it leaves his hand with an initial velocity of 97 mph to cross the plate about 0.4 seconds later with a velocity of 90 mph. We estimate the distance between the point where the ball begins its motion toward the plate and the point of release as 8 feet; that is, the pitcher holds the ball about 4 feet behind the pitcher's rubber as he begins to bring the ball forward toward his point of release, which is about 4 feet in front of the rubber. The released ball then has an energy of about $\frac{1}{6}$ horsepower-seconds. (A horsepower-second is equal to the energy output of a one-horsepower motor running for one second—and that is enough to lift 550 pounds one foot.) Making the crude—but useful—approximation that the force the pitcher applies to the ball is constant, we find that the throw takes about 0.11 seconds[8] and that the average force on the ball of about 12 pounds generates a mean acceleration of the ball of about 40 g's—forty times

---

[8]This is not the "delivery" time. The time between commitment—after which a balk would be called if the pitch were interrupted—and release is more like 0.8 seconds.

the acceleration of gravity. With these numbers, we find that the average power the pitcher transmits to the ball in the course of the pitch is about 1.5 horsepower! Since the pitcher's body is also put into motion by the contraction of his muscles (in fact, his hand and wrist are moving nearly as fast as the ball when it is released), we conclude that his musculature must have generated energy at a rate exceeding 3 horsepower during the action of throwing. Such power can only be generated by the large muscles of the thighs and thorax. As pitchers well know, it is important to develop and retain leg and body strength to throw the fast ball.

Although the flow of energy from the pitcher's body to the ball is surely complex and the description of that flow is not within the scope of this book, some comments about a particular aspect of the energy transfer are in order. In the course of the throw, the elbow leads the hand and ball at the midpoint of the action. At this time energy is stored in the stretching of the arm's tendons. That energy is transferred to the ball in the last portion of the throw as the spring-energy of the stretched tendons is released. When age or injury reduces the tendons' elasticity, the arm goes "dead."

In general, the elasticity of all tissues decreases with age, and the blazing fast ball of the young player must be replaced by the craft of the older pitcher. Late in his long and remarkable career,[9] Warren Spahn, commenting on his life in baseball, said, "When I was young and threw the fast ball. . . ." At 45, when he retired from baseball, Spahn was probably about as strong as he was at 25, but his arm was not as elastic.

PITCHING IN THE WIND

Excepting the recent intrusion of domed stadiums, baseball is an outdoor game and affected by the vagaries of the weather, which

---

[9]Spahn won over 20 or more games in a season 13 times; the last in a 23 and 7 season in 1963 when he was 42.

influences the mechanics of pitching and the facility of pitchers. Here we concern ourselves only with the weather's effect on the flight of the pitched ball, leaving the benefits of hot weather on pitching arms to physiologists and psychologists. Moreover, in the contest between pitcher and batter, we consider only the effects of weather on the pitched ball, leaving the problem of wind-blown fly balls and balls in the sun to outfielders and managers.

The weather affects the flight of the pitched ball almost solely through the velocity of the wind over the diamond. Though small effects follow from variations of temperature and barometric pressure—the fast ball is quicker by an inch or two on a hot day and at the lower air pressure before a storm or at a higher altitude ball park—these will hardly affect the game significantly. The average wind velocity over the United States is surprisingly constant at about 10 mph. (The wind surely blows harder at certain places, such as Mount Washington, but major league baseball games have not been regularly scheduled there.) Moreover, the prevailing wind is from the west and will usually be in the pitcher's face. Of course the weather is famously perverse everywhere—especially at Candlestick Park in San Francisco, where a balk was called on Stu Miller in the 1961 All-Star Game when the wind blew him off the mound—and winds can come, betimes, from any direction.

The wind's effect on the pitched ball is subtle and important to pitcher and batter. It is again convenient to consider specifically the effects of a 10-mph wind, though on the playing field of most major league parks the protection of the stands usually reduces the average wind velocity to a somewhat smaller level. The effects we discuss all vary almost linearly with wind velocity, hence, a light 5-mph breeze will have half the effect of the "standard" 10-mph wind.

As an example of the consequences of wind, we consider a standard major league fast ball, thrown overhand, that crosses home plate at a velocity of 90 mph 0.40 seconds after leaving the pitcher's hand with a muzzle velocity of 97 mph. We also assume the pitcher applied (a typical) backspin to the ball of 1800 rpm, which causes it to curve up and reach the plate perhaps 5 inches above where it would have without the spin. With the 10-mph wind in the pitcher's face, the ball will reach the plate slightly later (lagging by about 3 inches of flight path), traveling a bit slower at about 89.3 mph, and will arrive at about the same height. With the wind at the pitcher's back, the ball will reach the plate about 3 inches sooner at a velocity of 90.7 mph and cross the plate about an inch lower than it would without the wind. The ball will actually hop slightly less. These effects are all small compared with the precision of pitcher or batter, and not very important. But a crosswind can be a little less innocuous. If the wind is blowing across the field at a velocity of 10 mph, the fast ball will be blown about 3 inches—the diameter of the ball—off course at the plate, enough to trouble the nibblers who are trying to hit corners if the wind is erratic, as it often is. Again, the effects are roughly proportional to the wind velocity.

If the fast-ball pitchers can largely ignore moderate breezes, curve-ball and junk pitchers have a lot more trouble. We consider specifically a left-right deflection of a wide-breaking curve thrown so that it crosses the plate with a velocity of 61.2 mph in calm weather 0.6 seconds after leaving the pitcher's hand with a velocity of 70 mph. This particular ball, thrown with a sidespin of 1800 rpm, breaks about 15 inches. Thrown initially toward the inside of the plate, the ball will break over the outside corner. (Actually, the major league pitcher is usually more interested in the vertical deflection—down—than the transverse break.) With the wind directly behind him, or in his face, there is little significant change. The ball reaches the plate about 9 inches sooner (in flight distance) with the wind behind him and about 9 inches later with the wind in his face, but that is not important. The ball also breaks about the same amount, though here our lack of knowledge of balls near this speed leaves some uncertainty. A very slow curve may well break more when it is thrown against the wind.

A crosswind, however, modifies the pitch considerably. If the wind is from the north, from third to first, the right-handed pitcher's curve will break about 21 inches toward first base, but if the wind is from the south, the break will only be about 8 inches. Hence, a gusty day with crosswinds can cause serious problems with slow pitches. If the wind kicks up, a pitch thrown accurately at the corner of the plate can end up a ball or cross right over the center of the plate to be deposited in the bleachers by the batter as a souvenir for a fan.

TECHNICAL NOTE

a. the nominal similarity between the action of the table-tennis ball and the scuffed baseball may be a little misleading. Since the Reynolds number for the small ball, traveling at rather low velocities, is only about 10,000 (corresponding to a baseball velocity of only 10 mph), the separation of the boundary layer, important in the baseball knuckle ball and scuffed ball, is probably overshadowed for the table-tennis ball by the classical Bernoulli effect, that is, the air flowing over the Scotch tape has a longer path and must move faster, thus generating a smaller pressure that results in a deflection of the ball toward the tape. The similar effect of the speeding up of the air passing over the protuberant baseball stitches would not seem to generate sufficient force to account for the deflections of the heavier baseball.

BATTING THE BALL

A MODEL OF BATTING

There are many ways of batting the ball successfully through swinging at it so that precision of placement of the batted ball rather than extreme velocity is important. The drag bunt and the hit-and-run behind the runner at first base, for instance, represent purposeful hits that are made by swinging with less than full power. Hits to the opposite field often fall into this category. Conversely, the batter is often fooled by the pitcher so that his timing is thrown off and he swings weakly—albeit sometimes successfully hitting, perhaps, a "Texas leaguer" or an infield grounder that bounds so slowly that he beats the throw to first. In the following discussion, we do not consider the imperfect swings—perhaps a majority in the game—but only the full swing made with maximum effort. Only a few players—Babe Ruth and Ted Williams come to mind—*always* seemed to make full swings.

In considering the mechanics of the batting process, it is useful to construct a model of the full swing that is tractable but sufficiently close to a real swing by a real batter so that the consequences of the simple model illuminate the complex reality. Such modeling is simplified by the realization that a player swings a bat very much like a weight on the end of a rope; to a very good approximation, the forces on the bat exerted through the hands are directed along the axis of the bat. Hence, if the motion of the hands is known throughout the swing, the motion of the bat can be determined uniquely.

A model swing at a waist-high pitch by a strong player is illustrated in Figure 4.1, which shows a plausible trajectory of the bat throughout a swing in a horizontal plane calculated from

FIGURE 4.1   *The trajectory of the hands and bat during a typical swing. The position of the bat is shown at intervals of one fiftieth of a second.*

BATTER'S BOX

FIGURE 4.2 *The variation with time of the velocities (top), forces (center), and power to the bat and bat energy (bottom) during the model swing.*

simple dynamics and the trajectory of the batter's hands. The bat velocity is taken as the velocity of the radius of gyration with respect to the end of the bat which is near the vibrational node or "sweet-spot" of the bat—the best place to hit the ball. In the figure—and calculation—the initial position of the bat lies in the plane of the swing for convenience though most batters actually hold the bat more nearly upright. That difference does not seriously affect the consequent bat trajectory.

The graphs in Figure 4.2 show the variation with time during the swing of the velocity of the hands, the velocity of the barrel of the bat, the total force on the bat and the component of force in the direction of motion of the hands, the power supplied to the bat, and the kinetic energy of the bat.

During the swing, the batter moves his body toward the pitcher in "stepping into the pitch" and then rotates his body—hips, torso, and shoulders—as he brings the bat across the plate. Before the pitch, the batter usually stands with his weight placed over a point about halfway between his feet, which are spread to about the width of his shoulders; typically, the back foot is planted near the rear edge of the batter's box. When the ball approaches the plate, the batter generally shifts his weight backward an inch or two. This motion is halted, and reversed, by a large push from the rear foot—at which time there is almost no weight on the front foot. Then, the weight is shifted toward the front foot, usually in the course of a short step,[1] and when the bat crosses the plate there is almost no weight on the back foot. Typically, the batter's center-of-mass moves forward a little less than 18 inches in the one-fifth second required to swing the bat. The considerable energy of about 0.6 horsepower-seconds transferred to the bat in one fifth of a second (hence, an *average* energy transfer rate of 3 horsepower) is generated largely by the large muscles of the

---

[1] Babe Ruth often took two steps, beginning the first step about the time the ball left the pitcher's hand.

thighs and torso. The arms and hands serve mainly to transfer the energy of the body's rotational and transverse motion to the bat and add little extra energy to the bat.[2]

The swing can be separated usefully into two parts. For the first one tenth of a second, the character of the swing is dominated by the action of the batter; for the second tenth of a second until the ball is struck, the character of the swing is largely determined by the reaction of the bat. During that first part of the swing, the hands transmit a force of up to 50 pounds, largely generated by the rotation of the body, in pulling the bat about an arc. At the end of that time, the hands and the trailing bat reach a velocity of about 40 mph and the kinetic energy of the bat is about 0.2 horsepower-seconds. The weight shift during this time is equivalent to a motion of the body toward the pitcher with a velocity of about 8 mph corresponding to a kinetic energy of perhaps 0.7 horsepower-seconds. This energy is stored in the motion of the body.

After the one-tenth second elapses, the bat moves out of the arc of the hands and exerts an increasingly large reaction force on the hands and arms that will reach a value near 250 pounds as the bat crosses the plate. During this time, that large force straightens (uncocks) the wrists. The force from the reaction of the bat, transmitted through the hands and arms, slows down the body's forward motion (toward the pitcher) and the body's rotation; the body's stored kinetic energy is transferred to the bat,[3] increasing

---

[2]In particular, the contribution of the hands and wrists to the energy of the bat is almost negligible. Though the long-ball hitters—generally big, strong men—usually have large strong hands and wrists, there are exceptions. No National League player has ever hit more home runs in a season than Hack Wilson, 5'6" tall and 190 pounds. Wilson, however, built like a fire plug, wore size $5\frac{1}{2}$ shoes, and his small hands, "which matched his feet," could not have been especially strong. But the tremendous strength in his legs and torso was transferred efficiently to the ball through his strong arms.

[3]The part of the total force in the direction of motion of the hands is responsible for the energy transfer. That force ranges up to 150 pounds, requiring strong arms. This is why Pete Gray, the one-armed outfielder who played for the St. Louis Browns in 1945, hit no major league home runs. Using only one arm, he could not apply a force as great as the 150 pounds necessary to transfer enough of the energy developed in his strong thorax and legs to the bat so that he could hit the ball very far.

the bat's velocity from 40 mph to about 70 mph. The rate of that energy transfer exceeds 8 horsepower about 0.14 seconds after the initiation of the swing—and about 0.04 seconds before hitting the ball. This 8 horsepower is far greater than that actually generated by the contraction of muscles during this period. For many batters, the energy transfer from body to bat is nearly complete—the forward motion and rotation of the body stops almost completely when the bat crosses the plate.

Typically, the fast ball struck by the bat carries about one-fifth as much kinetic energy as the bat. If the ball is struck squarely, about half the energy of the swinging bat is transferred to the ball in the impact, so the speed of the bat is sharply reduced by about 30 percent. About one third of the original bat-and-ball energy is carried off as kinetic energy in the flight of the ball from the bat, and the rest of the energy (about three tenths of the original total energy) is lost in friction in the course of the distortion of the ball—and then to heating the ball.

The course of the swing after the ball is struck is kinematically irrelevant. The emphasis by coaches and sports teachers on the "follow-through" in baseball (and also in golf and tennis, for instance) is designed to insure proper actions *before* the ball is struck. Figure 4.1, however, does show typical positions of the bat and the hands after the ball is struck; the pattern would be a little different if the ball were missed.

The graphs in Figure 4.2 show a typical variation of the bat and hand velocities during the swing, the variation of the forces applied to the bat, and the energy transfers during this model swing. Though the precise quantities shown in the graphs derive from our particular model and must be quite special, the qualitative values of the quantities and the character of the changes of the quantities that are presented is probably close to that of real swings.

The batter in the diagram is assumed to have swung in such a manner as to drive a ball to center field as far as he can. If this right-handed batter[4] should miscalculate the velocity of the pitch and hit the ball 0.005 seconds early, and to right field, he would hit the ball before maximum bat velocity was realized and generally not hit the ball quite as hard. Since the loss of energy after the maximum is reached is usually small, if he swings too quickly, by 0.005 seconds, he will lose little power in driving the ball to left field.

A pull hitter swings so as to maximize the bat velocity a little later in the swing. In general, the pull hitter has more time and distance to apply force to the bat and can therefore transfer a little more energy to the bat and hit the ball a little harder. If the right-handed batter is to make a minimal adjustment in the position of the arc of his swing, he will naturally hit a ball over the inside portion of the plate later in the swing and to left field. He will hit a ball on the outside of the plate earlier in the swing and to right field. It is then harder to hit a very fast ball over the inside of the plate, and "high and tight" is a natural strikeout pitch for an exceptionally fast pitcher.

Aiming for the center-field bleachers, the batter in Figure 4.1 will swing the bat at a speed (of its prime hitting region) of about 70 mph so that the bat, passing through the hitting zone, is effectively rotating about a point near the handle end of the bat. If he swings as much as $\frac{1}{100}$ of a second early, the ball will go foul down the left-field line; if he is $\frac{1}{100}$ of a second late, the ball will go foul into the stands down the right-field line.

Batting is an intricate art, and there are various ways to bat

---

[4]Though the right leg bears more of a burden than the left for a batter who swings from the right side, the arms share the forces nearly equally. Hence, there is no natural advantage for a right-handed man in batting right-handed. Indeed, since the left-handed batter is favored in many ways, many players who throw right bat left: the batter who swings from the left side is closer to first base and moves in that direction as he finishes his swing, for some reason right-field fences tend to be closer than left-field fences, and there are more right-handed pitchers (whose slants are easier for a left-handed batter) than left-handed pitchers. There are, however, two famous first basemen who threw left and batted right, Hal Chase and George Bush. Chase, a great-fielding first baseman, was considered the best ever at the position by Walter Johnson and Babe Ruth while Bush, captain of a fine Yale team that went to the finals of the NCAA tournament, has done an exemplary job recently of throwing out the first ball on opening day.

well. If the batter in Figure 4.1 swings in such a manner (more of an "arm" swing) that his hands move faster through the hitting zone and the bat is effectively moving in a longer arc, perhaps as if it were rotating about a point about one-half a bat length beyond the end (toward the player), the bat will be rotating slower at the time of impact and the error of $\frac{1}{100}$ of a second, early or late, will result in a long fair ball to left or right field rather than a foul ball. With such a swing, less precision in timing is required, but the batter must begin the longer motion a little sooner than for the more compact arc of the "wrist" hitter.

Often, tall, strong hitters like Darryl Strawberry and Dave Winfield use a longer sweep and longer arc to balance their early swings (and better chance of misjudging the pitch) with the greater power generated in a longer swing and the less precise timing required in a sweeping swing. They also can afford to use a heavier bat. Smaller players, e.g., Howard Johnson, who hit with power more often tend to rely on quick, precisely timed swings with shorter arcs, as shown in Figure 4.1. They will choose a lighter bat to help them swing more quickly. With such swings, they can delay their commitment and better judge the ball.

Most distance hitters hold the bat at the end to maximize the bat velocity realized from the whip motion. Babe Ruth can be considered to hold the bat beyond the end! Batting left-handed, for much of his career he held the knob of the bat in the palm of his right hand.[5]

Players of yesteryear who emphasized precision, like Wee Willie Keeler, who succeeded by "hitting them where they ain't," often used long arcs and reduced the angular velocity. Ty Cobb and Honus Wagner separated their hands on the bat and swept the bat through the hitting zone, adding to the angular accuracy of their hitting (though Cobb, especially, would often bring his hands together early in the swing).

---

[5]Ruth, like most players, often made minor changes in his batting style. Pictures of the Babe on his follow-through after hitting his sixtieth home run off Tom Zachary in 1927 clearly show the knob of the bat below his right hand.

THE BALL-BAT COLLISION

For a well-hit ball, the bat-ball collision takes place near the *center of percussion* of the bat. An impact at this point on a free bat transfers no momentum—or force—to the handle. Conversely, a force at the handle is not transferred to the point of impact. Also, the collision with the ball takes place over a time of about $\frac{1}{1000}$ of a second, during which the contact point on the bat moves only about an inch and the handle moves a much smaller distance. The position of impact near the center of percussion, together with the small motion of the handle, means that at the moment the bat strikes the ball, the bat can be considered as a free piece of wood with the clamping effects of the grip of negligible importance. It is this property of the swing of the bat that allows us to divide the action neatly into the complex physiological actions that transfer energy from the body to the bat and the relatively simple processes that transfer energy from the bat to the ball. We address only the bat and ball interaction here.

Very large forces, reaching values as high as 8,000 pounds, are required to change the motion of the $5\frac{1}{8}$-ounce ball from a speed of 90 mph toward the plate to a speed of 110 mph toward the center-field bleachers in the $\frac{1}{1000}$ of a second of bat-ball contact. Hence, for a long home run, the force on the ball reaches a value near 8,000 pounds with an equal reactive force on the bat. Such forces distort bat and ball: The ball is compressed to about one half of its original diameter; the bat is compressed about one fiftieth as much. The collision is not elastic; much of the energy of ball and bat is dissipated in frictional heat.

Treating the bat as a moving free object with known weight, weight distribution, and velocity, and knowing the weight and velocity of the ball, one can calculate the initial velocity of the batted ball from the laws of mechanics given the inelasticity of the ball and bat. The ball may be considered as a spring; the bat applies force to the ball, compressing it, and the ball exerts force

FIGURE 4.3   *The distortion of balls and bats as a function of the applied force for the cycle of compression and expansion. The displacement here is the change in the diameter of the bats or balls.*

on the bat on regaining its original contours. The recoil from this exerted force propels the ball away from the bat.

Figure 4.3 shows the results of measurements by Paul Kirkpatrick of the force versus compression distance for a simulated bat-ball collision and similar diagrams for the compression of a golf ball and the compression of the bat. (Complete *stress-strain* measurements were not made on bats, and the curve shown for bats is an estimate presented to show graphically the small energy loss by the compression of the bat.) The upper curves of each set plot the force versus distortion distance for compression; the lower curves show the force versus distortion distance for the following expansion. The area under the upper curve is proportional to the energy absorbed by the ball in motion; the area under the lower curve is the energy returned by the ball in its resumption of its spherical shape. For an ideal spring, the two curves will coincide; the energy released from the spring on expansion is equal to the energy absorbed on compression. The area enclosed by the two curves is proportional to the energy dissipated or lost in friction. For the ideal spring this is zero. As shown here, in expansion, the baseball returns only about 35 percent of energy supplied in compression. The golf ball, much closer to a perfect spring, returns more than 75 percent of the compressive energy when struck by a driver.

The wooden bat is actually more elastic than the ball,[6] but since the wood is hard and the compression is small, the energy stored by the wooden bat is only about one fiftieth of that stored by the ball. If the wood were just as elastic as the ball, that energy which distorted the bat would be returned as efficiently as the energy of distortion of the ball, and the hardness of the bat would be irrelevant. If the wood making up the bat were so weak that the distortion was not returned at all, i.e., the ball would put a permanent dent in the bat, all that energy of distortion would be lost and the ball would not leave the bat with quite as much velocity.

Years ago, players used to hone their bats with a hambone because they believed this would make the bat harder and make the ball fly off the bat faster. Sometimes, batters would (illegally) hammer nails into the hitting area so the ball would strike iron. Since the ash bat stores so little energy on impact, and returns that energy more efficiently than the energy disposed in distorting the ball, hardening the bat would seem to serve no very useful purpose. Indeed, if the hitting area were armored with a thin sheet of absolutely hard and rigid steel, the well-hit ball would probably go slightly less far since energy would be lost as the distortion of the more elastic wood would be replaced by an increased distortion of the less elastic ball. But the scale of distance change is small at best. If the infinitely hard, armor-plated bat were substituted for the bats now in service, this would decrease the distance of a 400-foot home run only by about 2 feet.

---

[6]If you drop a ball and a bat, held vertically, an equal distance above concrete, the bat will bounce higher than the ball—and an aluminum bat will bounce much higher than a wooden bat.

By and large, the benefit from any effort by players to harden their bats—legally or illegally—must be considered largely psychological.

The inelasticity is usually described in terms of a *coefficient of restitution* (COR), which is the ratio of the velocity of the ball rebounding from the surface of a hard, immovable object to the incident velocity and equal to the square root of the proportion of the energy dissipated in the collision. For baseballs traveling 85 *ft/sec* (58 mph), striking a wall of ash boards backed by concrete, the mean COR of a large set of 1985 and 1987 official major league baseballs has been measured to be 0.563; the balls rebound with a velocity of 0.563×85 *ft/sec* = 48 *ft/sec*. Since the energy of the ball is proportional to the square of the velocity, the rebound energy of the 85 *ft/sec* ball is about $0.563 \times 0.563 = 0.32$ times the incident energy; the collision is quite inelastic; 68 percent of the energy is lost to friction (satisfactorily close to the 65 percent found in the static measurements of Kirkpatrick).

My colleague R. C. Larsen augmented the results at 58 mph commissioned by Major League Baseball and measured the COR of one National League ball to be 0.588 for an initial ball velocity of 25 mph and 0.584 at 18 mph. Isaac Newton noted through his experiments that the COR is nearly independent of velocity for uniform spheres. L. J. Briggs, however, found that the COR of a golf ball fell off considerably with increased collision velocity, and the COR for baseballs probably also decreases as the collision velocity increases. In the course of his measurements at the National Bureau of Standards of baseballs for the army in 1942, Briggs[7] found a value of COR of 0.46 at an incident velocity of 130 *ft/sec* (89 mph) for baseballs used in the major leagues in 1938. If these balls do not differ especially from today's balls, this indicates that the COR falls off sharply from the value of 0.563 measured at 58 mph to the value of 0.46 at 89 mph.

Of course, if the COR of the present ball is not much smaller

---

[7]Briggs, then Head of the National Bureau of Standards, published his results in the *NBS Journal* in 1945.

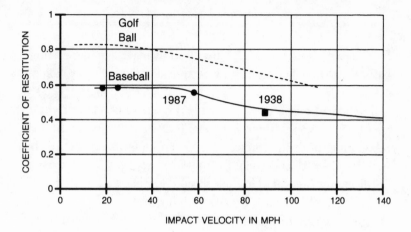

FIGURE 4.4 *Results of measurements of the coefficient of restitution (COR) of a 1987 baseball striking a rigid flat surface are shown by solid circles, and a value for a 1938 baseball is shown as a square. The solid line shows the estimate of the variation of the COR for present baseballs as a function of the impact velocity used in calculations. The broken curve shows the variation of the COR for a golf ball.*

FIGURE 4.5 *Plausible stress-strain cycles for a baseball struck by a bat for impact velocities of 58, 89, and 140 mph. The velocities are in the center-of-mass system (as a ball striking a stationary wall); the displacements are with respect to the center of gravity of the ball and correspond approximately to changes in the radius of the ball.*

at 89 mph than the measured values at 58 mph, the present ball might be assumed to be much livelier than the ball used in 1938. But this hardly seems likely; if the COR value of the present ball is as great as 0.55 for high impact velocities, it would seem that the 360-foot fly ball Lou Gehrig hit in 1938 would go 420 feet into the bleachers[8] today and that the 58 home runs Hank Greenberg hit in 1938 were with a dead ball! (Figure 4.12 shows the variation of the distance a ball can be hit with respect to the COR.)

Assuming that Briggs's measurements apply to today's balls, we can make an educated guess as to the possible variation of the COR with velocity that is shown in the graph of Figure 4.4. These values are used in the following calculations of velocities—and then distances traveled—of batted balls.

While the COR values for balls hitting a flat wall are not, in general, the same as for balls hitting a cylindrical bat, the bat values are probably similar to the wall values. Hence, the wall-ball values of the COR plotted in Figure 4.4, used for the bat-ball collisions considered here, probably represent reality to a satisfactory approximation.

With the insight into the stress-strain characteristics of the baseball provided by Kirkpatrick's measurements shown in Figure 4.3, together with the information concerning the coefficients of restitution shown in Figure 4.4, we can use the simple physical principles of the conservation of energy and momentum to construct a fairly reliable description of the ball-bat impact. The relevant information is contained almost completely in the model stress-strain diagram of Figure 4.5. Note that the "distortion" in Figure 4.5 refers to the displacement of the surface relative to the center of mass and is nearly the distortion of the radius of the ball.

---

[8]The striking difference between the COR of the 1987 balls, measured to be 0.563 at impact velocities of 58 mph, and the COR values of 0.46, measured for the 1938 balls at an impact velocity of 89 mph, can be understood in one of three ways, all unattractive: (1) the values of the COR do not vary strongly with impact velocity and the 1987 balls are much more lively than the 1938 balls, (2) the balls are about the same but the COR falls off sharply between 58 mph and 89 mph; or (3) one of the measurements is wrong (if so, Briggs's 1942 measurement must be in error). We choose (2) as the least unlikely.

FIGURE 4.6 *The top figure shows the variation with time of ball compression in a ball-bat impact. The lower figure shows the variation with time of the force between the ball and bat.*

Using the information depicted in the graph, we derive the variation of force with time and the variation of ball compression with time[a] during the ball-bat collision shown in the graphs of Figure 4.6. Note that although the bat-ball contact extends over about $\frac{1}{1000}$ of a second—a millisecond—the time of strong impact lasts only about one-half millisecond.

The strength of the impact can be defined usefully as proportional to the change in velocity of the ball. The velocity change corresponding to the impact velocity of 85 *ft/sec* (58 mph), at which major league baseballs are tested, is about 85 × 1.563 = 133 *ft/sec*, or 90 mph. This is about the same impact a baseball experiences when an 80 mph pitch is bunted—or when a stationary ball is given an initial velocity of 90 mph toward the outfield by a fungo bat. Briggs's measurements on the COR of 1938 baseballs at an impact velocity of 130 *ft/sec* (89 mph) concerned in a velocity change of about 190 *ft/sec*, which is about the change that occurs when a 200-foot Texas leaguer is hit off a 65-mph curve ball. But when a 400-foot home run is hit off a 90-mph fast ball, the velocity change of about 290 *ft/sec* corresponds to an impact velocity of about 140 mph. It is hardly surprising that the maximum forces on the ball, and maximum ball distortions, are very much greater for the home run off the fast ball than the bunt.

The diagrams in Figure 4.7 provide a somewhat simplified picture of the cross section of the ball at maximum distortion for

FIGURE 4.7   *Representative cross sections of balls at the maximum compression during the bat-ball impact for various impact velocities. The true distortions will be similar but can be expected to show more flattening on the side that meets the bat.*

the three impact conditions considered. The ball's elasticity for the small distortions that occur for bunts—or fungo hitting—is largely determined by the character of the outer yarn windings and is not much affected by the properties of the center core of the ball. Conversely, the elasticity of the ball that is very much distorted by the impact that results in the home run off the fast ball is strongly affected by the character of the core of the ball. The major league tests of the elasticity at velocity reversals characteristic of bunts are useful, but they do not establish the constancy of the ball's elasticity for the long home run.

THE SPIN OF THE BATTED BALL

For almost all bat-ball collisions, the line of momentum transfer through the ball does not coincide with the final direction of the ball's flight; the bat does not strike the ball perfectly square. The ball is not lined straight back to the pitcher but is hit to left or right field, on the ground, or in the air. From such an off-line impact, the ball will leave the bat spinning, though it was not spinning originally. Figure 4.8 suggests the origin of the spin of the ball induced by oblique collisions with a stationary flat surface such as a wall. Here the ball, traveling with an initial velocity v, strikes the wall at an angle θ. The initial ball velocity can be divided into two parts, or components, as suggested by the diagram: v_n is the component normal to the surface and v_t is the part tangential to the wall; v'_n and v'_t are the corresponding components of the rebound velocity v'.

We do not understand the stresses induced by the deformation of the complicated ball structure well enough to truly calculate the spin and trajectory of the ball as it rebounds from the wall, but we can deduce a sensible recipe that will allow us to make reasonable estimates of the state of the rebounding ball.

First, we can assume with confidence that the normal component of the rebound velocity is simply reduced from the incident

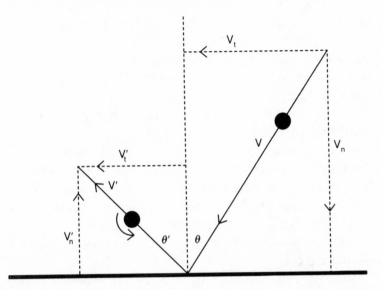

FIGURE 4.8   *Initial and rebound velocity components for a ball striking a flat surface at an angle. The curved arrow shows the direction of spin of the rebounding ball.*

normal velocity by a factor equal to the COR. For a value of the COR of 0.5, the normal velocity of the rebound v'_n would be just one half of the initial normal velocity v_n. If the ball struck a very slick wall obliquely and skidded off it, the transverse velocity v_t would be unchanged by the collision. But the forces between the ball and wall are so large for the collisions that interest us that we expect the ball to *roll* on the wall in the course of the collision just as if its surface stuck there. Then the ball would come off the wall spinning, and some of its transverse energy would go into the spin energy, reducing the transverse velocity of rebound. Moreover, this rolling of the highly deformed ball must take up some additional energy in internal friction. We would expect that this energy loss would be less than, but related to, the dissipative energy loss that reduces the normal component of the rebound velocity. If the value of the COR were near one and there were no frictional reduction of the normal rebound velocity, we would expect no reduction of the tangential velocity from frictional losses.

For the normal component of the velocity, which is reversed on impact, if the COR is 0.5 the rebound will be one half of the initial velocity. But the transverse component of the rebound velocity, which is not changed in direction, will retain about three fourths of the initial transverse velocity.

We calculate the spin of the batted ball struck obliquely by the bat using recipes that follow these considerations[b]. Although these kinematic relations refer ostensibly to the collisions of the ball with a stationary flat surface, within the uncertainties of the recipes, they should adequately describe the collisions of the ball with the cylindrical bat when properly translated to the system of the moving bat.

Measurements by Briggs, cited in his NBS *Journal* report, which show that any spin of the incident ball, i.e., from the pitcher, is largely removed in the course of the bat-ball impact are substantiated by the observations of players who speak of "straightening out the curve ball." Hence, although the fast ball, with its backspin, will come off the bat with a little less backspin and at an angle about one degree lower than the same ball thrown without spin—and the curve ball, thrown with overspin, will come off the bat with a little more backspin and about one degree higher—we do not consider these small effects in our model. (A degree up or down represents a foot up or down as the ball passes the pitcher or two feet as the ball sails past second base.)

In the explicit calculations, we assume implicitly that the bat is swung approximately in a horizontal plane.[9] Then the ball pulled by a right-handed batter toward the left-field foul line will strike the bat at an angle of about 20° from normal and the ball will be spinning counterclockwise as seen from above as it leaves the bat. If the bat strikes the ball below the ball's center, the fly ball will take off with backspin; if the bat strikes the ball over the

---

[9]Most batters direct the planes of their horizontal swings—at waist-high balls—about 10° upward to match the descent of the average pitched ball as it crosses the plate.

FIGURE 4.9 *The trajectory of a ball hit down the foul line showing the curvature produced by the sidespin induced by the bat.*

center, the ground ball that results will move across the infield with overspin. For a ball hit with a greatly tilted swing—such as the golflike swing at a pitch inside at the knees—the ball spin-axes are changed in an obvious fashion.

Since the spin of the ball affects its flight significantly, it is important to consider the spin induced by batting the ball in spite of the uncertainties in such spin calculations. Furthermore, even as there are uncertainties in our knowledge of the initial spin of batted balls, there are also uncertainties in our knowledge of the effect of the spin on the ball's flight (Chapter 2). Since the spin is important, however, we check the results of the models against that which we know from observations of the game.

As in pitching, the effects of up-down spin (usually backspin) are difficult to disentangle from the effects of gravity and then difficult to observe unequivocally. But the curves from sidespin are better defined. We thus calculate from our bat-ball recipe, and from our description of the Magnus effect in Chapter 2, the curve of a fly ball hit down the left-field foul line—over the left-field fence—and show the results in Figure 4.9. The ball carries an initial sidespin (clockwise as seen from below by the third baseman) of about 2000 rpm. We calculate that the ball, passing over the third baseman a few feet inside the base, curves about 25 feet foul at the 315-foot foul pole and lands in the street behind the ball park, after a carry of 380 feet, now "foul" by nearly 50 feet. Since this behavior is in rough accord with what we have all seen many times, we can conclude that our recipe is probably good within an uncertainty of, perhaps, 30 percent.

In these calculations, we assume that the motion of the bat's face is at right angles to that face and that all the sidespin of the ball follows from the ball's angle of incidence with the bat. Using the analogy of a golf shot, the club face is square with respect to its direction of motion. If this is the case, a ball hit to right or left field will always curve toward the foul line and a ball hit to center will have very little sidespin. Many batters, however, swing so the bat is moving along its axis when the ball is struck, giving a sidespin even to balls hit back toward the pitcher, in analogy to a slice or hook in golf. From watching the ball's flight, it seems that most batters hook a little—a right-handed batter hits the ball with a little extra counterclockwise spin (as seen from above), inciting a curve toward the left-field foul line. But some batters slice. Boston Red Sox catcher Rich Gedman, who hits with an "open" stance (a slice stance in golf), with the forward foot farther from the plate than the rear foot, seems to slice the ball. Hitting left-handed, Gedman's drives to right field may even curve toward the center fielder.

Using this recipe for the spin of the ball, we find that long home runs are typically hit so that the initial backspin of the ball is near 2000 rpm. Ground balls have similar rates of overspin that affect the bounce. The statement is sometimes made that such balls gain velocity as they bound through artificial turf infields. At 2000 rpm, the rotational velocity of the ball's surface at the spin equator is about 17 mph. A ball can only gain speed at a bounce, if that rotational surface velocity exceeds the regular linear velocity of the ball, which will be upward of 50 mph for balls hit hard enough to escape the infielders. Hence, ground balls will not speed up even with an overspin as great as 4000 rpm, but they will slow down less than balls with little overspin.

Rarely—but they are remembered in the nightmares of center fielders—line drives are hit to center field so that the ball has almost no spin, darts about like a knuckle ball, and is almost uncatchable. Balls hit to left or to right field will generally have enough sidespin to insure a smooth trajectory.

EFFECTS ON THE DISTANCE BALLS ARE HIT

Obviously, a ball travels farther if it is struck by a bat that is swung faster—or "harder." Less obviously, the fast ball hit by a full swing of the bat will travel faster and farther than a slow, change-of-pace pitch hit by the same speed swing. We will consider the magnitude of these effects.

For the sake of definiteness, in the following discussions we

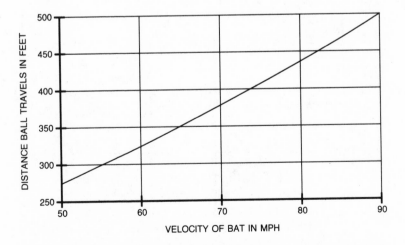

FIGURE 4.10   *The distance an 85 mph pitched ball can be hit by bats swung with different velocities.*

consider a bat, 35 inches long and weighing 32 ounces, striking a waist-high pitch so that the plane of the swing lies at 10° from the horizontal, i.e., the bat is swung slightly upward, to drive a ball at an angle of 35°. Among the variety of real swings, some vary considerably from this model swing; the swing at a ball passing over the inside corner of the plate at the knees often seems more like a golf swing. The salient points illustrated by the model, however, will be valid for all these swings.

Figure 4.10 shows the distance a ball that crosses the plate at a velocity of 85 mph travels when hit squarely with a full swing as a function of the bat speed. The assumption is made that the ball is hit at the center of percussion of the bat (a sweet-spot about $27\frac{1}{4}$ inches from the handle), and the quoted velocity is the velocity of that spot. Moreover, we assume that the ball is hit toward center field. A bat speed of about 77 mph is required to hit the ball 400 feet.

For a given bat speed, a solidly hit fast ball goes farther than a well-hit slow curve. Figure 4.11 shows the distance the ball

FIGURE 4.11   *The distance a pitched ball, traveling at different velocities, can be hit by a bat swung with a velocity of 70 mph.*

FIGURE 4.12   *The distance a batted ball can travel as a function of the value of the coefficient of restitution (COR) of the ball and bat if the bat velocity is 70 mph and the ball velocity is 85 mph. The* **X** *marks the estimated value of the COR for present balls and bats.*

travels for a bat being swung with a speed of 70 mph, striking balls that cross the plate at different speeds. The swing that hits a fungo (but with the bat described above, not a fungo bat) 300 feet will hit the 90 mph fast ball 370 feet!

Of course, a lively ball will travel farther than a dead ball. Figure 4.12 shows the distance a hard-hit home run ball will travel for different values of the COR under conditions such that the 90-mph fast ball will be hit 400 feet if the COR has the value of 0.430. Balls can be made that are much more lively and travel much farther. A baseball (wound with rubber like a golf ball) with a COR of 0.72 characteristic of golf balls would travel nearly 600 feet from the blow, which would have sent an ordinary ball 400 feet.

Major league specifications require the COR, measured at 58 mph, to lie between the values of 0.514 and 0.578, an allowable deviation of about ± 5 percent. If that proportional deviation holds for the smaller values of the COR that we believe are relevant at greater impact velocities, we might expect ball-to-ball deviations in flight length of the order of 10 percent. In fact, from measurements of 72 balls at an impact velocity of 58 mph, the mean ball-to-ball COR variation of balls now in use seems to be no larger than 0.005, corresponding to a deviation in distance of the 400-foot home run of only about four feet.

The sensitivity of the flight distance to the COR of a batted ball is suggested by Figure 4.12. The graph shows the flight distance of the ball as a function of the COR for balls hit under conditions that would drive a ball 355 feet if the COR were correctly given by the graph in Figure 4.4.

In Chapter 2, we noted that the conditions of storage of balls could affect the elasticity of the balls and then the distance balls could be hit—long flies hit with balls stored under conditions of extreme humidity could be expected to fall as much as 30 feet short of the distance expected for normal balls. The elasticity of balls stored under extremes of cold or heat can be affected also. There are apochryphal stories, going all the way back to John McGraw when the home team supplied the balls to the umpire one-by-one, of home team managers storing the balls to be given the umpires when the visiting team was at bat on ice. Those balls were taken off of the ice a few hours before game time so the cover would warm up and not alert the umpire but the core of the ball would remain cold and dead. Since McGraw managed, the rules have been changed so that *all* balls—still supplied by the home team—must be given to the umpires two hours before game time, hence any doctored ball will be used with equal probability by home and visitors alike rather evening things up. We have made some simple measurements[c] that suggest that the temperature effects can be substantial; deep-freezing to −10° F would seem to take about 25 feet off of a nominal 375-foot fly ball reducing many a home run to a loud out on the warning path. Such a ball, hit on the ground, would also skip through the infield a little more slowly. Conversely, storing the balls in a warming oven at 150°F, would seem to inject enough rabbit in the ball to take the 375-foot fly over the fence to land about 400 feet from home plate.

In general, it seems that a swing of the bat that would hit a ball that was at a temperature of 70°F, 375 feet, would drive a ball an extra 3 feet for every 10° increase in the temperature of the ball. Conversely, the ball would travel about 3 feet less for every 10° decrease in the temperature of the ball. Even in very cold days, the ball temperature probably seldom drops very much below 60° since the ball in play was probably stored at room temperature before the game and will be kept from cooling too much by the handling of the pitcher. It is more likely that the ball will reach the ambient temperature on a really hot day; if the temperature on the field is 100°, the balls probably reach that temperature also.

Though there are still stories of teams that rely on singles, speed, and defense storing all balls in a deep freeze until a few hours before game time when they play a slower team that depends more upon power and home runs, I am not a believer. Since the careers—and salaries—of the home team batters aren't helped much by the lower batting averages that must result from hitting a cold dead ball, any cold treatment of the balls must be kept secret; impractically secret to my mind.

It is interesting to note that a ball hit toward a foul line will go somewhat farther than a ball hit to dead center, though the bat speed is the same. This is because the ball must strike the bat at an angle (as shown in Figure 4.8) and the tangential part of the ball velocity v_t is not reversed by the bat-ball collision but retained with little loss in the blow. For the same bat speed, a batter will hit a 90-mph fast ball about 403 feet toward the left-center power alley with a swing that would hit the ball 400 to dead-center field. And if hit along the foul line, the ball would travel about 411 feet.

Although backspin adds distance to a long fly ball at a rate of about one foot in flight per 100 rpm of spin, there is probably little or nothing that a batter can do to change that spin. Any plausible batting stroke that takes a ball crossing the plate at a small downward angle of about 10° and propels it upward at the angle near

35° required to clear outfield fences will provide the ball with substantial backspin. Although the Ted Williams stroke that lofted Rip Sewell's Eephus pitch over the Fenway Park right-field fence in the 1946 All-Star Game probably generated almost no backspin, the Eephus pitch—tossed about 25 feet in the air to drop across the plate at a considerable angle—is seen today only in softball. Aside from losing the 20-foot flight advantage of backspin, Williams had to provide almost all the power—equal to an extra 50 feet or so of flight—to drive the 20-mph pitch over the fence. But only Babe Ruth could bat like Ted Williams, and no one else ever hit a home run off Sewell's blooper.

Batters sometimes work on their bats (illegally, but this is only a venial sin), grooving the surface to add backspin. It's unlikely, however, that there is any merit in this beyond occupational therapy. For all the small-angle impacts that lead to fair balls, the normal ball-bat friction is sufficient to ensure maximum backspin. Even as good, well-grooved tires do not stop a car faster than smooth, bald tires (as long as the car doesn't skid), fair balls hit off a smooth bat probably don't skid. Grooves, of course, are important on golf clubs, where hard steel, lubricated by the moisture from mashed grass, meets the smooth, enameled surface of the golf ball.

The inclination of the arc of the swing does not strongly affect the distance a ball is hit, but it does affect the mean angle at which the ball leaves the bat and, hence, the probability of hitting a very long ball and home run. The great high-average line-drive hitters such as Rod Carew and Wade Boggs swing the bat so that its barrel crosses the hitting region just in front of home plate, traveling upward on the same line that the average pitch is moving down, that is, at an angle of about 8 to 10°. (Players call this a level swing.) Then, if their bat position is correct but their timing is slightly off in an intent to hit the ball over second base, they will still hit the ball squarely. If they are early by 9 inches on the fast ball, as left-handed batters, they will pull the ball in a line drive between first and second into right field; if they are 9 inches late they will hit a line drive to the opposite (left) field.

FIGURE 4.13 *Ball and bat trajectories showing the consequences of a mis-timed swing by a left-handed batter with a level swing like Rod Carew and by a batter with an uppercut swing like Reggie Jackson.*

If their perfectly directed swing hits the ball absolutely square, that hit will be a line drive a few feet over the infielder's reach. If they hit the ball one-half inch high, they will hit a very hard ground ball that will often go through the infield; if they are one-half inch low, they will hit a high line drive for a double if it goes in the gap between fielders or down the line. But they will hit few home runs and very few pop flies. They will hit few home runs not because they do not hit the ball hard, but because they do not hit the ball up. Boggs, in particular, hits the ball very hard and wins home run contests against certified sluggers by swinging up a bit on the ball.

The relatively low batting average but high slugging average home run hitters like Reggie Jackson uppercut the ball, swinging typically such that the bat is moving upward at an angle of as much as 20° as it meets the ball. Roughly speaking, every 10° extra i.e., beyond 10°, of uppercut sends the squarely hit ball at an added upward angle of about 16° so Jackson's squarely hit ball takes off, going up at about 26° and Carew's solid hit comes off his bat, moving up at an angle of 10°. But for equally hard hit balls, Carew will end up sliding into second with a drive that landed about 210 feet from home plate and skipped between the out-fielders and Jackson will take a leisurely tour of the bases with the ball in the pocket of a fan who caught it 375 feet away in the right-field bleachers. If Jackson, the bigger man, actually hit the ball a little harder, it would go even farther.

But a mistimed swing will have different consequences for the two kinds of swing as suggested by the diagrams of Figure 4.13. If they miss their timing by swinging 9 inches too early on the fast ball, Carew will still hit the ball squarely but now on a line over the first baseman's head. However Jackson will top the ball and run out a weak grounder to the first baseman. If they swing equally late, Carew's squarely hit ball will pass in a line over the infield between short and third while Jackson, hitting under the ball, will hit a high pop fly to be gathered in by the third baseman. Hence, Reggie got the home runs while Rod hit with the high average. And both are on Bill James's list in his *Historical Baseball Abstract* of the 100 greatest who ever played the game.

BATTING AGAINST THE FAST BALL

The examination of the mechanics of the ball and bat's trajectory suggests the difficulties that face batters such as Ernest Lawrence Thayer's immortal "Casey at the Bat" in his attempt to hit the ball safely and win the game for Mudville. To put the accuracy of stroke required to hit a baseball successfully in a human context, we will examine the character, the precision, and the timing of the decisions a batter needs in facing a fast-ball pitcher.

For each 1.5 mph over 90 mph, the ball travels an extra distance of about one foot past the plate in a given time—the ball is a foot quicker—and drops less (or, subjectively, rises more) by about one inch. If the right-handed batter, aiming to hit the ball to dead center, misjudges the timing and hits the ball squarely, but one foot late, the ball will sail down the right-field line foul—or if he swings $\frac{7}{1000}$ of a second early (a foot early) he will hit the ball foul down the left-field line. If the batter aiming to hit a line drive misjudges the drop (rise) by 1.0 inches, he will hit a long fly ball—perhaps even a home run! If he misjudges the rise by 2 inches, the extra rise from an incremental velocity of $2\frac{1}{2}$ mph, he will foul the ball back into the stands behind the catcher. And if he swings at the Todd Worrell 95-mph fast ball at the height he would expect a fast ball traveling (only 91 mph), he will miss it altogether—even if he times it correctly.

Moreover, since the batter takes about one fifth of a second to swing his bat, he must start the swing when the fast ball is about halfway to the plate. He still has a little time to change his mind and reorient his swing—but not much. After his swing is under-way for one tenth of a second (and the ball is now about 15 feet from the plate), he won't be able to check it and he has little if any ability to change his point of aim. And about 50 percent of the deviation from curve, hop, or drop occurs in that last 15 feet.

And here we are considering hitting only the fast ball. If the batter sets himself for the fast ball on the three-and-one pitch, and the confident pitcher throws a curve or "change" instead with the same motion as the fast ball, the batter has serious adjustments to make in a very short time.[10]

In addition, to cover the 20-inch (17-inch plate width plus 3-inch ball diameter) strike zone, the batter must adjust his stride, and make his decision about the direction of that stride and weight shift very early in his swing. It is therefore extremely difficult for a batter to prepare himself to hit a fast ball effectively that passes either on the inside or outside corner of the plate. Most batters must implicitly give up one corner or the other—if not both—before the ball is even pitched.

Batting is difficult, and the Mudville fans must try to forgive "the mighty Casey" for striking out. (With two strikes on him, I have always felt that Casey should have choked up a bit and simply tried to meet the ball rather than swing for the fence.)

---

[10]Warren Spahn said, "Hitting is timing." Then he added, "Pitching is upsetting timing."

HOW FAR CAN A BALL BE HIT?

Stories of colossal clouts, of balls hit tremendous distances, consti-
tute part of the apocrypha of baseball. The stories of the longest
blows (550-foot carries are sometimes quoted) always seem to
place the action in some poorly surveyed situation, e.g., when the
estimate describes where the ball hit in the street outside the park.
Mickey Mantle is supposed to have hit a Chuck Stobbs pitch 565
feet out of Griffith Stadium[d] in 1953 and a Babe Ruth home run
hit in Tampa in the spring of 1919 was "measured" to have
traveled 600 feet. But some interesting, and better defined, nega-
tives are known—*no one*, not Ruth, not Gehrig, nor DiMaggio,
Mantle, or Maris ever hit a ball out of Yankee Stadium.

Although the question as to how far a baseball can actually be
hit must eventually be answered through experiment, we can
make some useful comments about the issue. First, though many
balls are hit 400 feet, such a blow constitutes quite a long home
run. According to Figure 4.10, the bat speed required to hit an
85 mph fast ball 400 feet to dead-center field under standard
conditions, such as no wind or moderate temperature, is about 76
mph. To hit the ball 450 feet, the bat must be swung at a velocity
of 86 mph, 13 percent faster with an energy 28 percent greater.
This is a very large difference, and we estimate that 450 feet is
the human limit *under such standard conditions*. Balls have been
caught about that far from home plate in the deep center fields
of the old Yankee Stadium and Polo Grounds. At Yankee Stadium, Al Gionfriddo caught a 420-foot drive off the bat of Joe
DiMaggio in the 1947 World Series, and DiMaggio pulled down
a Hank Greenberg drive in 1939 that was said to have traveled
nearly 450 feet. Willie Mays ran almost to the clubhouse stairs,
about 475 feet from home plate in the old Polo Grounds, to catch
a towering Vic Wertz smash with two men on to break the hearts
of the Cleveland Indians in the 1954 World Series. With the

hyperbole removed from a catch that didn't need it, the ball probably traveled about 450 feet.[11]

Although 450 feet may be the maximum distance a ball can be hit under normal conditions in Old New York, baseball is played under other conditions and other places. Let us say that the pitcher is very fast and the ball velocity is 95 mph. This increases the distance the ball would be hit from 450 to about 460 feet. If the batter pulls the ball toward the foul line, it will go as much as 15 feet farther, and we are up to 475 feet. Let us further assume that we are discussing a game played on a 100° day in July; then the ball could go 10 feet farther because the air is thinner, and another 10 feet farther because the hot ball is livelier, to land 495 feet from home plate. If a 10-mph following wind aids the flight, the ball could go another 40 feet, and we have a 535-foot blow. (We exclude meteorological freaks; the tornado that wafted Dorothy from Kansas to Oz could well have taken a baseball along.) In the thinner air in Atlanta—perhaps as the barometer is falling with lightning on the horizon—the ball might go another 10 feet to give us 545 feet. And if the foreman in the ball-manufacturing factory in Haiti had set the tension on the winding machine too high one afternoon so as to generate a gross of rabbit balls, who is to say how far one of *those* balls might go off a strong hitter's bat?

In summary, if someone tells me that a ball was hit 550 feet, I won't believe it—but I won't bet the farm against it either.[12]

---

[11]These balls would have stayed in the air about 6 seconds, enough time for a Mantle or Mays to run 150 feet under the ball.

[12]If one can possibly hit a ball 550 feet—maybe?—how far can one throw a ball? In 1957, at Omaha, Glen Gorbous threw a ball 446 feet. Assuming he threw on a hot day with a 10-mph following breeze at Omaha's altitude of 1,040 feet, we estimate that Gorbous must have thrown the ball with an initial velocity of about 120 mph. In Chapter 3, we noted that the very fastest pitchers throw with a "muzzle velocity" of about 110 mph. But pitchers throw from a stationary position and Gorbous threw from a run, which is sufficient to account for the extra velocity.

TECHNICAL NOTES

a. For a given velocity of impact, the area under the upper curve in Figure 4.5 is proportional to the initial impact energy, the area under the lower curve is proportional to the energy returned to the rebounding bat and ball, and the area enclosed by the curves is proportional to the energy dissipated in the distortion of the ball. The area under the force-time curve of Figure 4.6 is equal to the change in the velocity of the ball in appropriate units.

b. We use the following formulae in the course of calculations of the batted balls' trajectories:

$$v'_n = Cv_n \quad v'_t = \sqrt{\tfrac{5}{7}} \times \sqrt{0.5C + 0.5} \ and \ \omega = \frac{v'_t}{R}$$

where C is the COR, v_n and v'_n are initial and rebound normal velocities in the ball-bat center-of-mass system (which is equivalent to the ball-wall system of Figure 4.8), v_t and v'_t are the associated tangential velocities, ω is 2π times rate of spin of the ball (in revolutions per second), and R is the ball radius. The factor $\sqrt{\tfrac{5}{7}}$ accounts for the transfer of translational kinetic energy to rotational energy, and the factor $\sqrt{0.5C + 0.5}$ is introduced as an estimate of the reduction of the transverse rebound velocity that might follow from frictional effects. These recipes are designed to represent collisions at an angle less than 30° from the normal (which includes all fair hits). In the case of more nearly glancing collisions, the ball may skid off the bat, losing little transverse velocity and picking up little spin.

c. My colleague R. C. Larsen put several balls in a deep freeze held at about −10° F and several in an oven held at about 175° F for about a day. He then let the balls sit at room temperature for one hour and measured their coefficient of restitution (COR) at a velocity of 25 mph. As one might guess, the cold balls were dead—their COR was reduced about 10 percent from normal—and the heated balls were quite lively— their COR increased about 13 percent. This translates to a change in the distance of a fly ball hit 375 feet with a normal ball to about 350 feet for cold balls and over 400 feet for the hot balls.

d. According to legend, Mantle's drive carried 560 feet—but no one really knew where it landed. We know that the ball cleared the 55-foot-high bleacher wall, 460 feet from home plate, by about 5 feet where it glanced off a beer sign. Hence the ball was 60 feet in the air after it had traveled 460 feet. Mantle must have hit the ball upwards at an angle near 40° and balls always fall more steeply than their initial angle of projection upwards. If the angle of descent were 45°, the ball would have hit the ground after going 520 feet—60 feet past the wall. More detailed calculations suggest that it must have fallen at a slightly steeper angle and hit the ground about about 511 feet from home plate.

PROPERTIES OF BATS

THE OPTIMUM BAT WEIGHT

From the *Official Baseball Rules:*

1.09 (a) The bat shall be a smooth, rounded stick not more than $2\frac{3}{4}$ inches in diameter at the thickest part and not more than 42 inches in length. The bat shall be

(1) one piece of solid wood, or
(2) formed from a block of wood consisting of two or more pieces of wood bonded together[1]. . .

(b) Cupped Bats. An indentation in the end of the bat up to one inch in depth is permitted and may be no wider than two inches and no less than one inch in diameter. The indentation must be curved with no foreign substance added.

(c) The bat handle, for not more than 18 inches from the end, may be covered or treated with any material to improve the grip.

In fact, the bats used effectively by players are almost never longer than 36 inches. Once made of hickory and weighing from

---

[1]Any laminated bat design must be approved by the Rules Committee. In practice, laminated bats are not used.

40 ounces on up to the 56 ounces of lumber supposedly wielded by the young Ruth, today's bats are made of ash and range mostly from 31 to 36 ounces. The typical dugout bat rack in 1920 probably held no bat lighter than 36 ounces; the 1989 bat rack probably holds no bat heavier than 36 ounces. But then and now, the choice of weights and shapes varies considerably from player to player.

If we understand the mechanics of batting, it would seem we could establish the optimum size and weight of the bat from simple physical principles. A number of factors, however, suggest that it may not be possible to reach definite conclusions about the best size and weight for a bat: the broad choice of bats used successfully by baseball players tells us that proficiency in batting cannot depend sensitively on the bat's character; the game's sensitivity to small differences in batting means that precise analyses may be required if reliable conclusions are sought concerning those small differences (there is only about a 5 percent difference in the length of flight of the average home run and a long fly ball caught on the warning track), and the act of batting is, in itself, complex and not precisely understood.

At this time we will describe our analyses solely in terms of an optimum weight of the bat since this—along with the bat length—is a commonplace parameter. We neglect differences in the distribution of the weight along the bat, i.e., the shape of the bat, though that can be important; obviously, the addition of an ounce in the bat's handle has much less effect on the swing than the same ounce added to the barrel. We emphasize home runs because home runs are important and less dependent on tactical decisions than lesser blows.

The simplifying assumption that the weight of a bat can be changed without modifying its shape and weight distribution can only be exactly valid if a complete range of wood densities is available—and that is not the case. Baseball bats seem to have been made only of hickory or ash; the heavy bats used largely in the first quarter of this century were made of hickory with a density of about 0.82 times that of water; the light bats used today

are turned from ash with a density of about 0.638. A hickory bat
with the same dimensions as the 33-ounce ash bat Roger Maris
used to hit 61 home runs would weight about 42 ounces. The
48-ounce hickory bat Edd Roush swung (choked up) to give him
a lifetime batting average of .323 from 1913 to 1931 and the
47-ounce hickory bat Babe Ruth swung (held at the end) to hit
60 home runs in 154 games in 1927 (a record that still stands) had
larger barrels and thicker handles than Maris's bat and were
probably an inch or so longer.

The rules of baseball allow a barrel diameter of 2.75 inches. In
order to keep the bat weight down, the light bats used today
generally have smaller barrels—typically 2.50 inches in diameter,
which reduces the bat-ball hitting area (which is proportional to
the sum of the diameters of ball and bat) by about 4 percent and
the weight by about 4 ounces for an ash bat. The heavy bats
Mickey Mantle used and the 40-ounce bat Dick Allen swung had
full 2.75-inch barrels. Of course, longer bats weigh more; an inch
of extra length in the barrel adds nearly 3 ounces to the weight
of a hickory bat and a little more than 2 ounces to an ash bat,
though an inch added to the handle adds only about a third of
an ounce.

The great home run hitters have used a broad spectrum of bat
weights. Early in his career, when he led the American League
with 11 home runs in 1918, though he was used mainly as a
pitcher and spent only about one third of the season in the
outfield, Ruth was reputed to have used a 56-ounce hickory bat.
In 1927, when he hit 60 home runs, he seems to have favored a
47-ounce Hillerich & Bradsby slugger. Later he dropped to a
44-ounce bat, and was said to be experimenting with bats as light
as 36 ounces in his last year, 1935, with the Boston Braves—when
he quit early in the season with only 13 hits in 72 at bats, though
six of the hits were home runs. And Ruth's last home run, the last
he ever hit, was the first ball ever hit out of Forbes Field in
Pittsburgh!

But most players used much lighter bats then and use them
now. Roger Maris hit 61 home runs with a 33-ounce bat, and

Hank Aaron used 31- and 32-ounce bats to hit more major league home runs than anyone else. Though Ernie Lombardi used a 42-ounce bat in the thirties and forties, no other major batter since Ruth seems to have hefted more than 40 ounces.

If one player can hit a significant number of home runs with a 56-ounce bat and another player can be equally successful with a 32-ounce bat, one is led, inevitably, to the conclusion that the distance a well-hit ball travels cannot depend dramatically on the weight of the bat. Furthermore, since there are obvious advantages to a light bat as it must be easier for a batter to time a pitch with the quick swing possible from a light bat, there must be compensating advantages to a heavy bat or Ruth and Lombardi would not have used heavy bats. Though the heavier, and larger, bat, with its larger and longer barrel, provides more good hitting area, such a bat probably also transfers a higher velocity to the well-struck ball. It seems likely that a player can actually hit the ball a little farther with a heavy bat.

But could the players choose badly? Could an analysis of the physical principles of batting lead us in a different direction? Because of the difficulty of obtaining a full understanding of so complex an activity as batting, we must begin simply. From elementary principles of mechanics, we can say with complete reliability that for a given bat speed a heavy bat will drive a ball farther than a light bat. Conversely, for a given kinetic energy of the bat, a light bat will drive a ball farther than a heavy bat (for bat weights greater than 20 ounces). To this we add the (very plausible) condition that no player can swing a heavy bat faster than a light bat. And we hold that no player can put more energy into a light bat than a heavy bat. The energy transmitted to the bat is simply the product (better, *integral*) of the force applied to the bat through the hands of the batter times the distance through which his hands move. If we assume that the arc of the hands—and bat—is the same for light and heavy bats, the larger force that it is possible to apply to the more slowly moving heavy bat will result in a larger energy transfer to the heavy bat.

Figure 5.1 shows the variation of maximum distances with bat

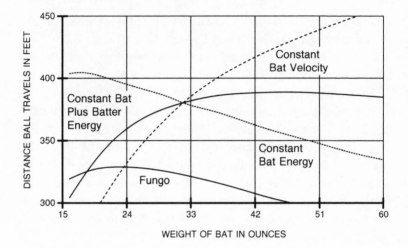

FIGURE 5.1 *The curves show the distances a 90 mph fast ball can be hit by bats of different weights swung at a definite velocity, swung with a definite energy, and swung so that the total bat-batter energy is held constant when a 90-mph ball is hit and when a stationary ball is hit (a fungo).*

weights that a 90-mph fast ball can be hit under these two conditions; the broken line shows the variation if the different-weight bats are swung with the same kinetic energy that drives a ball 381 feet from a 32-ounce bat; the dotted line shows the variation if the bats are swung with the same velocity as that of the 32-ounce bat. Noting that these curves establish *limits* on the variation of distance with bat weight, we can be reasonably confident that if a man who hits a ball 381 feet with a 32-ounce bat changes bat weight, and swings just as hard, the distance the ball will go will fall between these two extreme lines. In addition, it is clear that the man who hits 381-foot home runs with a 32-ounce bat is not likely to gain or lose very much distance by changing the weight of his bat by a few ounces.

Although this is as far as physics can carry us with—almost—complete reliability, it is interesting to investigate more detailed results from a simple, but plausible, model of bat and batter. In this model, we assume that both the bat and the body of the player move in the same general path, driven by the same muscular forces, though different-weight bats are used[a]. The forces generated by the muscles move the 180-pound batter just as they move the 2-pound bat. Since the same forces are applied over the same distances, the total energy generated to move bat and body during the swing is the same for different bat weights. If the bat is light, man and bat will move a little faster; if the bat is heavy, the swing will be a little slower. Fast or slow, light bat or heavy, the total kinetic energy of man and bat will be held constant, as we assume a constant muscular force and a constant path of motion through the swing. Though the energy is held constant for different bat weights, it is divided differently between bat and body. More of the energy goes into a heavy bat, though the bat—and player—velocity is smaller; less energy goes into a light bat, though the velocity of bat and player is somewhat greater. In the case illustrated by the solid line in Figure 5.2, the kinetic

FIGURE 5.2 *The trajectories of balls struck by a bat swung under the ball by given amounts. The bat speed is 70 mph directed upward by 10°; the ball speed is 85 mph directed down by 10°. The diagram in the upper right shows the configuration of the bat and ball when the bat is swung under the ball by three-fourths of an inch.*

energy of the 32-ounce bat is about the same as the kinetic energy of the 180-pound ballplayer. We again choose the total energy as that which would drive a 90-mph fast ball an optimum distance of 381 feet.

The real batter-bat interaction is much more complicated than the model. In fact, energy is exchanged between bat and batter in an intricate fashion. Moreover, physiologists note that the forces put out by muscles decrease as the velocity of contraction of the muscles increases, suggesting that less force is applied generating less energy when the muscles move more quickly in swinging the lighter bat. Hence, the model is not meant to describe the batter-bat interaction in detail but to simulate the results of the real interaction in some useful approximation.

The model is certainly simplistic, and the conclusions derived from it must be taken lightly. But the results that suggest the maximum distance of 389 feet would be achieved if the batter would use a 46-ounce bat are interesting and not implausible. The effects are not large: the 46-ounce bat propels the ball only 8 feet farther than the 32-ounce bat; if the batter drills a hole in his 32-ounce bat (perhaps filling it with cork) so the weight is reduced to 28 ounces, the ball will go only 373 feet.

Armed with the model (albeit simplistic), one can look at other factors. How does the optimum bat weight vary with the size of the player? Babe Ruth was reported to weigh 251 pounds in 1927 when he hit 60 home runs. According to our model, the Babe would have been best suited to hit 90-mph fast balls for distance with a 52-ounce bat. Conversely, Lloyd Waner probably weighed little more than 140 pounds at times. Our model tells us that "Little Poison" would hit fast balls hardest with a 42-ounce bat.

Using the approximation that the total energy output of a ballplayer is proportional to his weight, we can estimate the dependence of the length a player can hit a ball on his size. Hence, we find that the same proportionate effort that allowed the 140-pound Waner to drive a 90-mph fast ball 338 feet with a 32-ounce bat would allow 180-pound Ducky Medwick to hit the ball 362 feet using the same-weight bat. And if 225-pound Mark McGwire swung equally hard, his shot could land in the bleachers 382 feet from home plate.

If the ball is moving more slowly, the optimum bat weight declines. The 180-pound nonpareil model player would best use a 48-ounce bat to reverse the velocity of a 95-mph Roger Clemens fast ball and send it toward the center-field fence. But a 42.5-ounce bat would better drive the 80-mph slider, and the 65-mph breaking ball would be hit farthest by a 37-ounce bat. And if the batter wanted to hit fungoes (or bat a ball off a tee), he would find that a 24-ounce bat would be best—if he could find one. Indeed, special fungo bats are made to be quite light by reducing the diameter of the barrel.

By and large, these results are in accord with experience and common sense. Except for the fungo bat, however, all of the optimum bat weights seem rather high, even if they are optimized for slower ball velocities than the major league fast ball. But to survive in professional baseball, one must hit the ball often as well as hard. The model player would lose only a little distance by going to a 32-ounce bat, and with the extra quickness of swinging the lighter bat, he could expect to make good contact more often.

Have any measurements been made that might substantiate the results of the simple model that heavy bats drive the ball farther? Yes! Many years ago, *This Week* magazine (May 20, 1962) reported the results of an experiment in which Roger Maris batted for distance with five different bats that varied from 33 ounces to 47 ounces. The bats were copies from Hillerich & Bradsby records of the bats of previous great home run hitters (Ruth used the 47-ounce bat to hit his 60 home runs). The pitcher was a veteran Yankee batting-practice pitcher who expertly served up hittable pitches. Though there were fluctuations in the length of the drives, of course, a statistical analysis confirmed the trend seen casually: the heavier the bat, the farther Maris hit the ball. Indeed, the best fit to the data showed an increase of range with bat weight about like that given by the dotted constant-velocity line of Figure 5.1, though that limit was within the uncertainties of the data. Maris probably adjusted his swing so that he swung the heavier bats almost as fast as the lighter bats; perhaps he lengthened his swing when he used the heavier bats to hit the batting-practice pitching.

However, although the average length of balls he hit with his own 33-ounce bat was the shortest of all, Maris returned to that bat to face hostile pitching.

If the loss in distance that follows from a choice of a lighter bat is not great, is the gain in time in swinging the lighter bat substantial? It is easy to set limits on the difference in timing using the extreme model of the constant-energy bat. If we assume that the "time of decision" is typically 0.15 seconds before the ball crosses the plate, we can conclude with some assurance that the gain in the time required to swing the bat that follows from the change from a very heavy 38-ounce bat to a moderately light 32-ounce bat can be no greater than 0.0133 seconds, or about 1'9", in the flight of the 90-mph fast ball. Our more realistic model cuts that down to about 1'3". The batter would gain a little over a foot on the fast ball by shifting from a 38-ounce bludgeon to his favorite 32-ounce bat. If he drills a hole in that 32-ounce bat, reducing the weight in the barrel by about $1\frac{1}{2}$ ounces, he would gain about 0.005 seconds, or about 6 inches, on the fast ball. For some players, the small gain in time may be more important than the small loss in the speed—and flight distance—of the ball as it comes off the slightly lighter bat.

Such a simplistic, essentially kinematic, argument ignores the athletic complexity of the swing of the bat. The fine batter's "beautiful" swing that supplies maximum energy to the bat as it strikes the ball follows from precise timing of the various actions that contribute to the swing. It seems possible that for a given batter, a slightly lighter—or slightly heavier—bat may lead to a better (timed) swing and a larger energy transfer to the bat. And this could well override any pedantic kinematic derivation of an ideal bat weight.

Though these discussions define the considerations applicable to choosing an optimum bat weight—and support the common-sense view that players now choose bat weights sensibly—is it not possible that a better weight distribution would serve to drive a ball faster and farther? Could the standard bat be badly designed?

The kinetic energy the player supplies to the bat can be divided into two parts: an energy associated with the linear motion toward the ball by the mass of the bat and an energy associated with the bat's rotation about the point of impact with the ball. This rotational energy, which is typically about 5 percent of the total energy, does not contribute toward driving the ball. If a bat could be designed so that almost all of the mass were concentrated very near the point of impact (a bat with a short, heavy barrel and a handle like the wooden shafts of Bobby Jones's golf clubs of the 1920s would do the trick), the rotational portion of the bat energy would be much reduced. Such a bat, swung with the same total energy, would hit the ball somewhat farther; the ball hit 400 feet with the standard bat would go almost 410 feet from the golf club–bat. But the small increment of extra distance would be achieved at the cost of a considerable reduction of the region of good hitting and a considerable increase in the cost of broken bats. A ball hit a few inches from the optimum point along the axis of the bat would probably not leave the batting box much faster than the broken-off head of the bat.

HITTING THE BALL IMPERFECTLY

This discussion of batting has so far only considered almost perfectly hit balls. Since it is the pitcher's goal to see that balls are not well hit—and pitchers are sufficiently successful to hold mean batting averages well below 30 percent—miss-hit balls are the rule rather than the exception. The bat must be swung with great accuracy with respect to the trajectory of the pitched ball if the ball is to be well hit. We have noted that the swing must be timed correctly with an error no larger than $\frac{1}{100}$ of a second if the ball is to be hit fairly between first and third base.

Rather precise vertical orientation is also required. Figure 5.2

shows the trajectories of balls hit by a bat with a 2.75-inch-diameter barrel with different bat-ball vertical displacements. We assume that the swing is otherwise well directed and the bat is swung with a speed of 70 mph upward at an angle of 10° at an 85 mph overhand fast ball, which crosses the plate dropping at an angle near 10°. The ball that is hit squarely is driven 200 feet as a line drive, rising to a maximum height of about 13 feet. If the bat is swung under the ball by 2 inches so that the line of motion of the bat's center passes under the ball's center by 2 inches, the ball will go onto the roof of the stands behind the catcher. The ball hit one inch below center will be a routine fly ball out unless it is hit down the foul line, where it might be a home run. The ball will go farthest—about 350 feet—if hit about $\frac{3}{4}$ of an inch below center. If the bat has a smaller 2.50-inch barrel, as is the case for most of the light bats used today, the characteristic offsets must be reduced by about 4 percent.

The paths of ground balls are more difficult to show, but, excepting Baltimore chops, where the ball rebounds from home plate so high that the batter is on first before it comes down, the ground ball that goes for a base hit must not be hit too sharply into the ground. By and large, for a ball to be hit so that it evades the infielders, the bat should strike the ball no lower than $\frac{3}{4}$ of an inch below center and not higher than $\frac{3}{8}$ of an inch above center.

The batter is permitted a larger error along the bat's axis. An error of 2 inches is quite permissible, and if other criteria are satisfied, he can hit the ball well, though the impact is more than 3 inches along the axis from the optimum point on the bat.

Though the nominal latitude of error of the position of the ball-bat impact point along the bat's axis is greater than the allowable up-down error, this batter can adjust to changes in the ball's vertical position more easily than changes in its horizontal position. He does not need to change his stride or body orientation sharply to deal with the allowed vertical range of strike pitches, but he must change the position of his body to cover the 20-inch width of the strike zone. (The strike zone is one ball diameter wider than the plate since the umpire is supposed to call

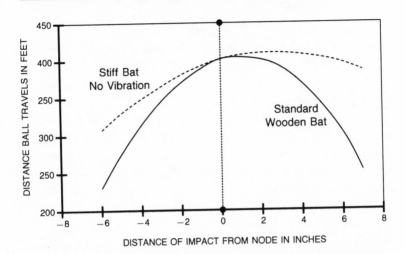

FIGURE 5.3   *The variation of the distance a ball can be hit as a function of the distance of bat-ball impact from a vibrational nodal point of the bat.*

a pitch a strike if any part of the ball passes through the strike zone.)

If a bat strikes a ball at a point on the bat at a distance from an optimum point too near the handle or end of the bat, the velocity of the ball from the bat is reduced. For expositional simplicity, it is interesting to describe this loss of velocity in terms of a loss of distance in a long drive to the outfield. The velocity variation with position along the bat axis is determined by simple kinematic criteria, defining the loss of energy from collision-induced rotation of the bat and the loss of energy to vibrations of the bat along its length.

The broken curve in Figure 5.3 shows the variation of distance the ball travels as the point of impact varies along the bat if the bat is (unrealistically) so stiff that oscillations can be neglected and (unrealistically) one can neglect the clamping of the end of the bat by the hands of the batter.[b] For the particular bat under consideration, which is 35 inches long, the zero point is chosen to be a point (the center of percussion) about $27\frac{3}{8}$ inches from the handle. The bat handle will be driven forward into the left palm of the right-handed hitter by bat-ball impacts toward the end of the bat beyond the center of percussion; the handle will be driven backward into the right palm by impacts nearer the handle; impacts at the center of percussion will not affect the motion of the handle. The force exerted on the hands by a miss-hit with a stiff bat is not great, not nearly enough to sting the hands. The ball's impact—which would sting the bare hands if caught—is greatly reduced by transmission through the bat. The maximum energy transfer from the stiff bat to the ball occurs when the ball is struck at a point about 30 inches from the handle or about 5 inches from the end of the bat. Energy is lost in the impact of balls hit at the end of the bat, or near the handle, by the increased rotational energy the impact transfers to the bat. Hence, the variation in distance, reflecting a variation in velocity of the driven ball, follows simply from kinematic criteria.

But bats vibrate when the ball is hit too far from the optimal point, a vibrational node (which is a point of no vibration), resulting in a weakly hit ball and often a broken bat. And that vibration does sting the hands. Under the large forces the ball's impact applies to the bat, a bat's vibrational mode becomes excited, as suggested by the diagrams in Figure 5.4. The diagram at the left shows the bat striking the ball at a node; the bat is not distorted. The next two diagrams show the distortion of the bat (center-line) on striking the ball at vibrational antinodes—points of maximum vibration—near the end of the bat and near the handle. (The distortions are exaggerated in the diagrams; the maximum excursions are less than one-half inch.) The diagram at the far right suggests the envelope of the vibrations that result from miss-hits.

The magnitude of the induced vibration is proportional to the natural amplitude of vibration at the point of impact; hence, no vibrations are set up when the ball strikes the bat at a node and maximal vibrations are set up if the point of impact is near an antinode. The energy of vibration is proportional to the square of the amplitude of vibration—twice the amplitude, four times the energy.

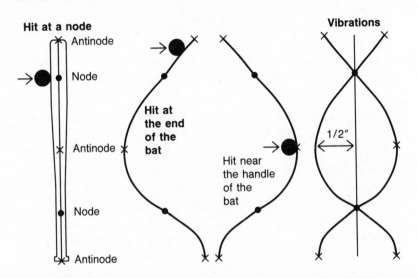

FIGURE 5.4 *Distortions of a bat on striking a ball at a node, at the antinode at the end of the bat, and at the antinode near the center of the bat. The diagram at the right suggests the character of the vibrations induced by hits away from the node.*

Since bats are not regular geometric objects, it is difficult to calculate the vibrational properties of a particular bat and harder yet to generate a recipe useful for all bats. It is possible, however, to make intelligent estimates of the magnitude of the vibrational effects (with the assistance of *dimensional analysis*). The amplitude of the vibration induced by the collision of the bat and ball must be proportional to the impulse (the force applied to the bat multiplied by the time over which the force is applied), inversely proportional to the weight of the bat, and inversely proportional to the frequency of vibration. Knowing the impulse to the bat, and the weight of the bat, and having measured the frequency of vibration of the bat, a typical amplitude of vibration induced in the bat by striking a ball near an antinode is determined to a little less than one-half inch; the bat will be distorted about as much as the ball. Using the same kind of estimate, it seems that a

significant fraction of the bat-ball collision energy can be transferred to the bat's vibrational energy.

For the specific, though typical, bat considered here, the frequency of the oscillation was measured to be 260 cycles per
second (middle C). Since the corresponding characteristic half-
cycle time of about 0.002 seconds is appreciably longer than the
natural bat-ball interaction time of about 0.0005 seconds (see
Figure 4.6), the bat does not return the energy of distortion to
the ball but retains that energy in the vibration familiar to any
baseball player. That energy is, of course, lost to the ball, hence,
the ball will rebound with decreased velocity.

The solid curve of Figure 5.3 presents a rough estimate of the
distance a batted ball travels as a function of the distance of the
impact from the optimum point (which is near a node of
the vibration) when the energy lost to the vibration is taken into
account.

Although the half period of vibration is sufficiently long compared to the time the ball is in contact with the bat so that the
energy is not returned elastically to the ball, the act of deforming
the bat does lengthen the time of bat-ball contact (as the bat gives
way at the ball's impact). The highest-frequency sound produced
in the course of ball-bat contact is roughly one-half the inverse of
the impact time (so that a collision that lasts less than $\frac{1}{1000}$ of a
second generates sound frequencies greater than 500 cycles per
second, which is one octave above middle C). One hears these
characteristic high frequencies in the "crack" of the bat. When
the ball is hit off-center, the collision time is a little longer,
generating somewhat lower maximum frequencies. With a lower
maximum frequency and the addition of a strong component of
lower-frequency sound from the natural bat oscillation, the
"crack" becomes more of a "thunk." Outfielders use the sound
as a part of the pattern to which they respond. A one-time center
fielder told me that when the ball is hit straight at you (the
distance such balls will travel is difficult to judge) if you hear the
bat "crack," run back (toward the fence); if the sound is a
"thunk," run in (toward second).

It is interesting to note that the heavier, thicker-handled bats of yesteryear were more rigid and had higher natural frequencies and smaller amplitudes of vibration upon off-center impacts with the ball. Since hickory has about twice the elastic modulus of ash, the stiffer hickory bats vibrated with a smaller amplitude and higher frequency. (Though, for a given configuration, the vibration frequency varies inversely with the square root of the density, and hickory is denser than ash, this reduction of the natural frequency from the greater density of hickory is not as important as the increase in frequency from the greater stiffness of the hickory.) Since the energy stored—and then lost—in the vibration of the bat is proportional to the square of the amplitude, these stiffer bats hit the ball more effectively when the point of impact is far from a node.

Since the stiffness of a beam is determined largely by the stiffness at the nodes where the maximum bending takes place, the effective stiffness of the whole bat is largely determined by the stiffness at the node on the handle (as shown in Figure 5.4), typically about 6 inches from the knob. In turn, that stiffness depends strongly (about as the cube) on the diameter of the handle; a change from a 1-inch diameter handle to a $1\frac{1}{4}$-inch diameter almost doubles the stiffness of the bat, increasing the length of the good-hitting region. The extra thickness of the handle will add about an ounce and a half to the weight of the bat, but since that weight will be in the handle the "feel" of the bat will not be much changed. The additional weight at the handle end of the bat may also help broaden the good hitting zone by increasing the rotational inertia somewhat, thus making the bat more difficult to turn because of the impact of the ball away from the center of percussion. This effect decreases that loss of energy to bat rotation by such a miss-hit ball. Although the increase in bat mass will slow the swing down somewhat, that effect is probably small. Hence, if a player is comfortable using a thicker-handled bat, he might gain two or three hits over a season on inside pitches and add a few points to his batting average.

In general, bats with thicker handles and bats with longer barrels have longer vibration-free zones of good hitting. Edd Roush, hitting choked up with his 48-ounce thick-handled hickory bat, probably never in his life stung his hands hitting an inside pitch. The long-barreled "bottle bats" used by players such as Heinie Groh[2] and Bucky Dent make sense—for some players.

ABERRANT BATS, ALUMINUM BATS

Aluminum bats are now used in baseball played outside the major leagues. Bats constructed in strange and ingenious ways are used in baseball variants, such as various forms of recreational softball, where there seem to be no special constraints on bats. The use of illegally altered bats have occasionally been detected in major league baseball. Questions arise concerning the efficacy of these aberrant bats.

The thin-handled bats made of light wood that are popular today are fragile and not cheap. Hence, the cost of bats for amateur baseball and minor league baseball is significant on the economic scale of these activities. Since bats made from aluminum tubing do not break and can be made with appropriate balance and weight, they have become the bats of choice for such baseball and are now admitted by the rules. It is now hard to find a wooden bat among the rows of aluminum bats in most sporting goods stores.

The requirement that wooden bats be turned from one piece

---

[2]Groh's bats had a barrel about 17 inches long, which necked down abruptly to a rather thin handle also about 17 inches long; this odd bat was about half barrel and half handle. Groh swung the 46-ounce bat choked up with his hands slightly apart. Groh had small hands, and he preferred a thinner handle than was common in the 1920s. Dent used a more conventional bat but with a long, full barrel that he swung choked up. Groh and Dent were largely line-drive hitters—though each hit important home runs. Indeed, Groh led the 1919 Cincinnati Reds—who beat the Chicago Black Sox in the World Series—in home runs (with five) and in slugging average (at .431).

Recently, some have used the name "bottle-bat" differently—for any thick-handled bat.

of solid wood places constraints on the size of light bats. Since woods less dense than ash are not strong enough to serve as material for bats, a light bat must perforce contain less wood and be smaller than a heavier bat. Moreover, the weight distribution along the bat is defined by the bat's shape, and the traditional shape is that which makes the best use of wood.

Since the thickness of the aluminum tubing used to make the hollow aluminum bats can be varied over wide limits, the weight and weight distributions of aluminum bats can be set almost independently of the size and shape of the bat. Presently, aluminum bats are shaped like traditional wooden bats and are made to have similar weights and weight distributions[c], but this follows from convention rather than as a mechanical necessity. The freedom of design afforded by aluminum has, however, resulted in the manufacture of light bats with the full-size 2.75-inch barrels allowed by the rules of baseball, and wooden bats must have smaller barrels to keep to the same weight. Hence, the typical 32-ounce wooden bat will have a barrel diameter of only 2.50 inches. With the larger barrel, the 32-ounce aluminum bat will have about a 4 percent larger effective bat-ball hitting surface with no compensating disadvantage.

Aluminum bats may also be better than wooden bats in other ways. The question has been raised as to whether a ball hit by an aluminum bat will go faster and farther than a ball hit by a wooden bat. Players also hold the view that an aluminum bat has a longer region of good hitting than a wooden bat. In particular, balls hit short on the bat—near the handle—are propelled more efficiently by an aluminum bat, and with less vibration and stinging of the hands, than by a wooden bat. There appears to be some consensus among players that these reputed advantages of the aluminum bat are real. Are their conclusions correct? If so, why?

We address first the question of the length balls are hit by an aluminum bat (or the speed with which ground balls hit by an aluminum bat travel through the infield). We have noted that the baseball is not very elastic. On striking a hard surface, only about

30 percent of the energy is returned in the rebound. The baseball is more elastic than a beanbag, but not by much. The coefficient of restitution for a beanbag striking a hard surface is near zero; beanbags do not bounce at all. No matter how strong the batter is, a baseball bat will not hit a beanbag much past second base. Nevertheless, it is possible to design an implement that will hit a beanbag a long way. An examination of such implements can provide some insight into mechanisms that allow baseballs to be hit farther by bats.

Though we will not be able to hit the beanbag very far with a baseball bat, we will do better with a tennis racket faced with a sheet of very thin rubber instead of gut or nylon strings. With this contraption, in the collision of the beanbag with the racket, the beanbag will travel past the plane of the racket by perhaps a foot, stretching the rubber. Then the rubber will react, propelling the beanbag away from the racket with a rather high velocity in a kind of catapult effect. Actually, a sheet of rubber would induce too much air resistance for a really fast swing of the racket-bat, and we would do better by stringing the racket with very strong, but quite elastic, strings.

If we attach a long handle to the tennis racket and weight the racket head to copy the length and weight distribution of a baseball bat, and then swing the racket like a bat, we will be able to hit a beanbag the size and weight of a baseball quite a long way. If we are able to find strings as elastic as gut or nylon that are sufficiently strong so they will not snap under the large forces we would generate, we could hit the beanbag farther than a baseball struck by a bat! (We could probably find steel springs that would do the job by replacing the conventional gut stringing of the racket.) Or we could hit a baseball farther with our racket-bat than with a regular wooden bat.

We explain this phenomenon by describing the (imaginary) observation of the results of a collision between a specific (also imaginary) racket-bat and the beanbag-ball (a baseball cover stuffed with beans so as to have the same weight as an official ball):

The beanbag was flung toward the plate by a major league

fast-ball pitcher and the racket-bat was swung by a major league power hitter. When the beanbag hit the racket face, the strings gave way elastically and the bag moved 6 inches past the plane of the racket face. (This deflection is about ten times the compression of the baseball hit by a bat.) Then the strings reacted elastically, flinging the beanbag back toward the center-field bleachers.

The bat-beanbag collision took place in about $\frac{1}{100}$ of a second (rather than the $\frac{1}{1000}$ of a second for the collision of the wooden bat with the baseball), and the maximum force was 800 pounds (rather than the 8,000-pound force that reversed the baseball flight). The 800-pound force compressed the soft beanbag-ball one inch and distorted the racket-bat by 6 inches, as we noted. Then about six sevenths of the collision energy was stored by the racket and only about one seventh by the beanbag. Since the beanbag is completely inelastic, all the beanbag compression energy was lost to friction, but only 40 percent of the energy stored by the racket was lost. Then, for the collision of beanbag and racket, more than 50 percent of the collision energy was returned to the beanbag as kinetic energy in the catapultlike action of the racket on the beanbag, corresponding to a COR greater than 0.700. For a baseball struck by a wooden bat the energy return is only about 32 percent, corresponding to a COR of about 0.565. A racket-bat swing, which would drive a baseball 400 feet using a wooden bat, would drive the beanbag about 480 feet (and a baseball a few feet farther). The racket-bat would hit a beanbag farther than the wooden bat would hit a baseball! And if the racket-bat were used on a baseball in Yankee Stadium, the ball could well be knocked clear out of the stadium—and not even Babe Ruth ever hit a ball out of Yankee Stadium.

Even an umpire as obtuse as fans claim in their calumny would be suspicious of such a bat. But aluminum bats seem to have a little of the catapultlike elastic properties of the racket-bat and probably can hit a baseball appreciably farther than a wooden bat. From measurements made by my colleague R. C. Larsen on the compressibility of aluminum bats, it seems that, for a given force, the distortion of an aluminum bat is about one-tenth[d] as great as the distortion of the ball (rather than the factor one-fiftieth for wooden bats). Moreover, from the high frequency and persistence of the sound emitted by the bat when struck lightly at a hitting point (at a node of the longitudinal vibration), it appeared that the distortion was quite elastic and the compression energy would be restored quickly (unlike the longitudinal vibration). Hence, in this mode the bat was surely much more elastic than the ball while the relatively incompressible wooden bat stores little energy and is only slightly more elastic than the ball. Then, the aluminum bat stores about one-eleventh of the collision energy in a highly elastic deformation of the bat, which was returned efficiently to the ball, and the ball stored about ten-elevenths of the energy in the deformation of the ball, most of which was lost in friction.

Though the intrinsic elasticity of the bat will be near 100 percent, some of the bat deformation energy will be retained in the energy of motion of the aluminum shell that makes up the bat. We estimate that 80 percent may be returned to the ball. The elasticity of the ball at this impact velocity will be about 19 percent, which is the square of the COR at the relevant impact force. Using these values, 24.5 percent of the collision energy will be restored to the bat and ball, a value much greater than the 19 percent for collision of the wooden bat and ball. This corresponds to an effective COR of 0.50 rather than the value of 0.43 appropriate for the wooden bat-ball impact at that impact velocity, and the home run hit 380 feet with the wooden bat will go as much as 415 feet from the aluminum bat. The same kind of calculations suggest that a fungo hit 330 feet by a (regular) wooden bat would travel about 12 feet farther hit with the same swing from a similar aluminum bat.

Though the numerical values presented here can be considered only as a kind of illustrative estimate—in view of our uncertain understanding of the character of the aluminum bat—the aluminum bats now used for baseball will surely hit a ball appreciably farther than a wooden bat.

The second purported advantage of the aluminum bat is that it is more effective than wood on inside pitches hit too near the handle. The aluminum bat is a kind of shaped hollow cylinder; since aluminum is about four times as dense as ash, to achieve the same weight the bat must be hollow. Aside from the greater strength of the aluminum, the hollow cylinder is more rigid than a solid structure containing the same mass of material; typically, the aluminum bat is about twice as stiff as the wooden bat—twice as much force is required to bend it a given amount. Hence, when a ball is struck badly—near an antinode at the handle or the end of the bat—the stiffer aluminum bat takes up much less energy in longitudinal vibration than the wooden bat—and stings the hands less. Moreover, since the vibrational frequency is higher[e] than for a wooden bat, the aluminum bat even returns some of that energy to the ball. Hence, the response of the aluminum bat to a ball hit away from the vibrational node (or sweet-spot) is close to the dotted curve of the perfectly rigid bat shown in Figure 5.3. A ball miss-hit, near the handle or the end of the bat, will go farther off an aluminum bat than a wooden bat. There is more room for error with an aluminum bat.

ABERRANT BATS, CORKED BATS

On occasion, players have modified their wooden bats by drilling an axial hole in the end of the bat and filling it with elements such as cork, cork balls, and rubber. They place a wooden cap over the hole and sand and varnish it to hide the modification. Neither the hole nor the addition of the filling are allowed under official baseball rules.

It is desirable to determine whether such an illegal modification might create a bat with properties that a legal bat cannot achieve—and hence threaten the integrity of the game—or if the effect of the modification can only produce a bat with properties no different than might be achieved with a legal bat design.

It is not possible to conclude categorically that a specific physical change in a bat is good or bad. But it is possible to determine the character of the change and to determine if the change—if illegal—would produce a bat with characteristics outside the range of bats constructed according to baseball rules.

The properties of a bat relevant to striking a ball squarely are largely defined by three weight distributions, or three moments. We list these together with representative values (taken from a particular bat that was 35 inches long and weighed 32 ounces). All weights are expressed here in ounces, and lengths are expressed in inches.

a. The sum of the weight of each piece of the bat, which is just the total weight of the bat (the zero moment) [32 ounces].

b. The sum of the weight times distance, measured from the handle, for each piece of the bat (the first moment) [765 inch ounces].

c. The sum of the weight times the square of this distance for each piece of the bat (the second moment or moment of inertia) [21,000 square-inch ounces].

There are three key positions along the bat that follow from the moments—defined here as lengths measured from the handle end of the bat.

d. The center of gravity [24 inches].

e. The center of inertia [25.6 inches].

f. The center of percussion (a sweet-spot) [27.3 inches].

There are other factors of interest:

g. The liveliness or elasticity of the bat.

h. The resonant frequency of the bat [260 cycles per second].

i. The position of the vibrational node [27.35 inches].

Though the nodal point and center of percussion will be close for most bats, the near equivalence for this particular bat is an accident.

For a given bat, the center of gravity (d) is just the point of balance of the bat—as balanced on a finger, for example. The center of percussion (f) can be found by holding the bat lightly at the handle end, e.g., by finger and thumb, and striking the barrel with a light hammer. When the bat is struck at the center of percussion, one feels no motion at the handle. For most bats, this point is very near the vibrational node (i); when the bat is struck by the hammer at the node, no vibrations will be felt at the handle end. The center of inertia (e) can also be determined primitively but less simply. If the bat is laid out on a smooth waxed wooden floor (or better, on an ice surface) and struck or pushed vigorously at a point so that the bat begins to slide without rotating appreciably (in the plane of the surface), that point is the center of inertia.

The moments are physically manifest in straightforward ways. The weight (a) is felt when one simply holds the bat vertically. The force required to hold the stationary bat out horizontally is proportional to the first moment (b). The force required to wave it back and forth vigorously when it is vertical is proportional to the second moment (c). This second moment contributes most to the "feel" of the bat and is the factor most important to the batter.

The elasticity (g) is determined by the bat's resilience near the point of the bat that strikes the ball; a resilient bat may store energy upon impact and return that energy to the ball. Wooden bats are not very resilient—and thus store little energy. The resonant frequency (h) is a measure of the energy loss when a ball is struck away from the vibrational node. A high frequency indicates a larger, i.e., longer, sweet-spot. Bats with long barrels or thick handles will display higher vibrational frequencies and larger sweet-spots.

An interesting illegal bat modification is made by drilling an axial hole in the end of a bat and filling it with an extraneous material (e.g., cork, cork balls, rubber, rubber balls). We first consider the effects of the hole—without filling.

As an exercise, I have been able to show explicitly that for a specific illegal modification of a specific bat, the changes in the moments (a, b, and c) and the characteristic lengths (d, e, and f) that define the bat's character can be achieved through legal changes. The changes will not affect the elasticity (g). While it is possible to invent extreme illegal corklike modifications that could not be copied legally, e.g., making the whole end of the bat cork, such bats would probably not be much good in hitting a baseball.

We compared the characteristics of a specific bat before and after a specific illegal modification and after certain legal modifications. The bat was an inexpensive "Louisville Slugger–Wade Boggs Model" purchased from a local sporting goods store. It was 35 inches long and weighed 32 ounces. The illegal modification was chosen to be a hole $1\frac{3}{16}$ inches in diameter drilled 6 inches deep along the axis from the barrel end of the bat. The hole was filled with cork. The density of the wood was 0.638—a typical density for American ash (or white ash)—and the density of the cork was 0.25 times the density of water.

Such a modification does sensibly change the bat. The weight is reduced by about 1.5 ounces, and that loss is wholly at the end of the bat. If a player is having trouble getting around on the fast ball, this weight reduction will help him in very much the same way that a lighter bat might help: the batter will probably not drive the fast ball as far by 2 or 3 feet when he hits it well with the drilled-out bat, but he may hit it with good timing more often.

Aside from eliminating the hollow sound from the air column created by the hole in the bat (such a sound might excite an excess of interest by the umpire), the cork or rubber stuffed into a hole drilled in the bat will serve more as a detriment than an advantage. The extra material will add, perhaps, one-half ounce to the bat's weight—at the end—and then store about 2 percent of the bat's energy. But that energy will not be effectively transferred to the ball. Even if the filler were quite elastic, the elastic energy of the filler cannot be transferred efficiently to the bat in the $\frac{1}{1000}$ second of the bat-ball collision. Hence, the extra material will only slow the bat down a little and slightly reduce the distance the bat hits the ball. Such a filler would thus take about 3 feet more off a 400-foot drive.

Whatever the advantages of the lighter bat, it was possible to show that very near the same results achieved by the illegal modifications could be achieved legally by (a) choking up on the bat by about one inch, (b) using a bat with the same dimensions but constructed of slightly lighter wood, (c) sawing (legally) about $\frac{3}{4}$ of an inch off the end of the original bat, and (d) legally turning the barrel down from 2.50 inches to 2.40 inches. Both the modified bat (d) with the reduced barrel size and a bat shortened by one-half inch and made of slightly lighter wood fit the moments and characteristic lengths of the illegally drilled bat almost exactly.

Solution (b) implies that a satisfactory "lighter" wood is available, which may not be the case. The American ash from which bats are made has an unusually high strength-to-weight ratio. Ash was celebrated in medieval times as the only proper wood from which to construct the lances of knights errant; an ash lance was light enough to carry and wield and strong enough to impale the opposition. Any reduction in the weight of the wood seems to require a substantial compromise in strength and hardness. But Green (or Pennsylvania) ash or black walnut, for example, might provide adequate lighter—if somewhat weaker—bats.

We have noted that aluminum bats are much more resilient than wooden bats and upon striking a ball store energy in compression, which is returned to the ball efficiently. It is conceivable that a device could be fashioned to insert into a hole drilled in a bat that could store energy upon impact, which would then be returned efficiently to the ball (in much the same way that the flexion of an aluminum bat stores, and returns, energy). Such a device could improve the effectiveness of a bat, but some reflection suggests that it would be difficult to design—and quite difficult, if not impossible, to make so that it would be hard to detect.

The resonant frequency and position of the vibrational node are only important for miss-hit balls. Nevertheless, such illegal holes drilled in the bat have no significant effect on either that frequency or the node's position.

In summary: The characteristics of any specific baseball bat can be changed by drilling an axial hole in the end of the bat and filling it with some light, inactive, extraneous material. The modified bat differs from the original bat by its lighter weight and smaller moment of inertia. Bats drilled out in this way are excluded from play under the *Official Baseball Rules.* But the properties of such modified bats can generally be reproduced by a legally constructed bat. In particular, a slightly shorter bat made of the same-density wood, or a bat with a slightly smaller barrel, or the same-size bat made of somewhat less dense wood will have nearly the same "feel" and hitting characteristics as the bat lightened by having a section of the core replaced by a lighter material.

The only substantial caveat to this conclusion follows from the limitations on properties of wood. If the batter demands a lighter bat *exactly* the size of his regular bat—and will not tolerate a modest reduction in hardness and strength—he can best achieve that result by illegally drilling a hole in it.

TECHNICAL NOTES

a. In this model, we calculate the maximum velocity v of the bat, from the relation

$$v = \sqrt{\frac{2E}{m + \epsilon^2 M}}$$

where m is the mass of the bat, M the mass of the player, and E is a normalization constant approximately equal to the total energy produced by the muscles. The constant, ϵ, is taken as equal to $\frac{1}{6}$, which is in accord with the observation that the maximum energy of bat and player are roughly equal. Using this relation, for a given total energy E, we determine the variation of v with m for batters of a given weight

M. Then, as discussed earlier, we determine the velocity that a pitched ball will travel when struck by the bat of mass m moving with the velocity v.

In the approximation that the ratio of muscle mass to total mass is about the same for players of different size, we can expect that the energy put out by the muscles is proportional to the weight of the player. (For example, if the shape of a small player were the same as a player 30 percent heavier, the muscles of the larger player, 10 percent longer and with a 20 percent larger cross section, would be expected to exert a 20 percent greater force over a distance 10 percent greater to develop 30 percent more energy.) Hence, we can set E proportional to M and estimate how much farther a large player might be expected to hit a ball than a smaller player.

b. The approximation that the bat can be treated as a free piece of lumber, which is adequate for balls hit by the bat near the center of percussion (effectively the sweet-spot), may break down for balls hit too near the handle or the end of the bat. The clamping action of the hands restricts the rotation of a (unrealistically) stiff bat reducing the energy loss to such rotation. Hence, the broken curve must overestimate the energy loss of a stiff bat—if anyone can find such a bat—for impact points far from the sweet-spot. This suggests that a firm grip on the bat might help for balls hit on the end or the handle of the bat—while a loose grip would do as well for a well-hit ball. But that advantage of a firm grip is reduced if the bat is as flexible as most wooden bats are. For a real bat, the signal of the impact travels along the bat as a transverse vibrational wave at a velocity that varies along the bat—very fast at the thick barrel, more slowly near the thin handle. We can estimate the velocity of the signal by assuming that the typical 20 inch ($1\frac{2}{3}$ feet) distance from the node at the barrel and the antinode at the handle is about one-half wavelength. Using the measured vibrational frequency of 260 cps, the signal speed will then be about $2 \cdot (1\frac{2}{3}) \cdot 260 = 867$ ft/sec and the signal will take about 2 milliseconds to reach the hands from the point of bat-ball impact. With the duration of that impact of the order of $\frac{1}{2000}$ of a second, the handle and hands will not receive the signal—and respond to the impact—until the ball has left the bat.

c. In fact, manufacturers seem to make aluminum bats with somewhat different weight distributions than wooden bats. Since the weight of a section of the hollow aluminum bat is in the shell or skin of the bat and the mass of the similar section of the solid wooden bat is distributed throughout the whole section, there is a tendency to place relatively less weight in the barrel of the aluminum bat and more in the handle. If the same gauge (or thickness) of aluminum were used throughout the bat, a one-inch section of a 2.75-inch-diameter barrel would weigh about 2.75 times as much as a one-inch section of the handle. For a wooden bat of the same dimensions, the ratio would be much greater—about $(2.75/1.0)^2 = 7.56$. Usually, the manufacturer only partially counters this by using thicker material in the barrel than in the handle of the aluminum bat.

The different weight distribution of the typical aluminum bat has both advantages and disadvantages. With less weight in the barrel behind the impact with the ball, the 32-ounce aluminum bat acts in the collision more like a 30-ounce wooden bat, transfers less energy to the ball and loses some distance on a solid hit. But the trampoline action of the bat acts to drive the ball much farther than a wooden bat anyway, and the effectively lighter bat can be swung a little quicker. Also, since the extra weight in the handle increases the turning moment of the bat, a ball miss-hit too near the end of the bat or its handle will lose less energy in turning the bat. But the reduction of vibration by the stronger handle is more important in increasing the good hitting region.

d. The measurements of the compressibility of the balls and bats were made statically and can only approximate the actual forces generated by the collision of the spherical ball with the cylindrical bat. We believe the results more likely underestimate, than overestimate, the bat values.

e. When an aluminum bat is struck sharply away from the nodal sweet-spot, the bat will ring with two frequencies. There will be a very brief high-frequency note and a persistent lower-frequency sound. The higher note is the frequency of the longitudinal vibration, the lower note is generated by the organ pipe–like excitation of the air column in the bat and has a half wavelength equal to the length of the column (a little more than 30 inches) and a frequency of about 220 cycles per second.

INDEX